Guidelines for small-scale fruit and vegetable processors

by
Peter Fellows
Midway Technology Ltd.
St. Oswalds Barn, Clifford
Hay on Wye, Hereford
United Kingdom

Published by arrangement with the
Food and Agriculture Organization of the United Nations
by
Daya Publishing House®
A Division of
Astral International Pvt. Ltd.
New Delhi - 110 002

"This book was originally published by the Food and Agriculture Organization of the United Nations (FAO) as FAO Agricultural Services Bulletin No. 127".

"The designations employed and the presentation of material in this publication do not imply the expression of any opinion, whatsoever on the part of the Food and Agriculture Organization of the United Nations concerning the legal status of any country, territory, city or area or of its authorities, or concerning the delimitations of its frontiers or boundaries."

'The designations 'developed' and 'developing' economies are intended for statistical convenience and do not necessarily express a judgment about the stage reached by a particular country, territory or area in the development process.'

First Indian Edition, 2007
Reprinted, 2024
ISBN: 978-81-7035-501-4

Published by : **Daya Publishing House®**
A Division of
Astral International Pvt. Ltd.
– ISO 9001:2008 Certified Company –
4760-61/23, Ansari Road, Darya Ganj
New Delhi-110 002
Ph. 011-43549197, 23278134
E-mail: info@astralint.com
Website: www.astralint.com

Printed at : Replika Press Pvt. Ltd.

Preface and Acknowledgements

This book is intended to assist planners and field workers who are involved in the promotion of small scale fruit and vegetable processing in developing countries. It sets out the different reasons why food processing may be important: for improved nutrition and food security; for employment and income generation. It notes the different approaches to project development that are needed to meet one or more of these differing objectives.

Many entrepreneurs may also find the information contained in the book helpful for practical implementation of the different aspects that are needed to ensure a successful business. These are not simply concerned with producing a high quality food, but also include methods of business planning, market research, securing agreements with suppliers and retailers, and monitoring the finances of a business to ensure a profitable operation.

Much of the information has come from fieldwork that I have been involved with in Africa, Asia and other regions during the last twenty years and the approach to writing the book has been very much to record my experiences and those of colleagues, rather than trying to compile yet another academic text. As such some professionals and experts in their own specific subject areas may take issue with a perceived lack of rigour in my treatment of their topic. This may or may not be justified, but I have attempted to treat each subject in a way that is accessible to, and understandable by an intelligent non-specialist; I would welcome all comments on the text, particularly if you feel that this could be better achieved.

I would like to thank all of the very many entrepreneurs and field-workers who have given so freely of their time and expertise during our meetings. Their detailed knowledge of how to practise small scale fruit and vegetable processing under difficult conditions, especially in rural areas and their unbounded enthusiasm, often in the face of enormous obstacles to success, have been inspirational. Without their assistance, this book could never have been written.

I would also like to acknowledge the following for permission to use their information: Midway Technology for Figures 1, 4, 25, 43 and 44, Intermediate Technology Publications for Figure 2, Table 23 and the text in Appendix III, FAO for Table 27, UV Systems plc, Sudbury, Suffolk, UK., for Figure 38 and Mike Dillon Associates for Figure 53.

My grateful thanks also to Mathew Whitton, for the line drawings, done in his own inimitable style, to Pamela Hopkinson for editing the text and not least to Wendy Bullar, for her ideas and constructive criticism, her unending patience and support during the preparation of the manuscript.

Glossary and Abbreviations

Acid preserves	foods that have a high acid content that inhibits spoilage.
Adulteration	deliberate contamination of foods with materials of low quality.
Average weight	a known proportion of containers have a fill-weight above that shown system on the label.
Brix	units of measurement of sugar concentration
Case hardening	the formation of a dry skin on a wet food due to over-rapid drying. It slows the rate of drying and can lead to spoilage during storage.
Chlorination	the addition of chlorine to water to destroy micro-organisms.
Contamination	materials that are accidentally included with a food (e.g. dirt, leaves, stalks etc.).
Critical control	stages in a process where quality control can have a major effect on
Points	food quality.
Cross	the transfer of soils or micro-organisms from raw food to contamination processed food.
Enzymes	natural proteins in foods that can cause changes to colour, flavour or texture of the food.
Equilibrium Relative Humidity	the moisture content at which a food does not gain or lose weight and is stable during storage
Fill-weight	the amount of food placed into a container or package and written on the label (also net weight).
HACCP (Hazard Analysis Critical Control Point) system	a system used to identify and control contamination in food processing.
Humidity	the amount of water vapour in air.

Hydrometer	an instrument that measures specific gravity of liquids, used to measure salt, sugar or alcohol concentration.
Limited liability	the liability of shareholders in a company to the nominal value of their shares
Low-acid foods	foods that have little acid and therefore can contain food poisoning bacteria if poorly processed.
Micro-organisms	tiny forms of life, invisible until they are in large numbers, including moulds, bacteria and yeasts.
Minimum weight	all packages have a fill-weight equal to or greater than that shown on the label.
Net weight	the amount of food filled into a container.
Osmosis	movement of water through cell walls of plants or micro-organisms
Pectin	a natural gelling agent found in some fruits
pH	a scale used to express acidity or alkalinity, from 1 (strong acid) through 7 (neutral) to 14 (strong alkali).
Potable water	water that will not cause illness.
Preservation index	a figure that is calculated to show that the amounts of acid, sugar and salt are enough to prevent spoilage of foods.
Quality assurance	a management system which controls each stage of food production from raw material harvest to final consumption.
Quality control	a series of checks and control measures that ensure that a uniform quality food is produced.
Refractometer	an instrument that measures the refractive index of a liquid, which is used to measure soluble solids in syrups or salt in brines.
Shelf life	the time that a processed food can be stored before changes in colour, flavour, texture or the number of micro-organisms make it unacceptable.

Sodium benzoate	a chemical preservative that is particularly effective against yeasts.
Sodium metabisulphite	a chemical preservative that is effective against moulds and yeasts.
Soils	any material that contaminates food or equipment (e.g. grease, scale, burned-on food or other food residues).
Un-incorporated	a group of people formed into an association that does not constitute a association single legal entity.

Abbreviations

ASEAN	Association of South East Asian Nations
CCP	Critical Control Point
HACCP	Hazard Analysis Critical Control Point
ROPP	Roll on Pilfer Proof (caps for bottles)
TOTO	Twist on Twist off (lids for jars)
UHT	Ultra-High Temperature (sterilisation of foods)
UNDP	United Nations Development Programme
USA	United States of America.

Table of Contents

		Page
Preface and Acknowledgements		iii
Glossary and Abbreviations		v
Abbreviations		ix
I	GENERAL INTRODUCTION	1
PART 1	**PROCESSING FOR HOME CONSUMPTION**	7
1.1	**Introduction**	7
1.2	**Food security, nutrition and health**	8
	1.2.1 Improvements to home processing and storage	10
	Drying	10
	Concentration by boiling	12
	Fermentation	12
	Pickling	14
	Storage	15
	1.2.2. Home processing to earn extra family income	16
PART 2	**PROCESSING FOR SALE**	19
2.1	**Introduction**	19
2.2	**Selecting products and production methods**	25
	2.2.1 Fried products	25
	Additional processing notes	25
	2.2.2 Bottled and canned products	25
	Additional processing notes	25
	2.2.3 Dried fruits and vegetables	28
	Additional processing notes: blanching, sulphuring and sulphiting	30
	Syrup pre-treatment	33
	Types of dryers	34
	Packaging	35
	2.2.4 Chutneys, pickles and salted vegetables	37
	Chutneys	37
	Additional processing notes	38
	Pickles	39
	Salted vegetables	41
	Additional processing notes	41

Page

2.2.5 Pectin and papain 43
Pectin 43
Papain 44
2.2.6 Sauces 45
Additional processing notes 45
2.2.7 Juices 47
Additional processing notes 47
Acidity 47
Pulping 48
Pasteurization 48
Filling 50
2.2.8 Squashes, cordials and syrups 50
Squashes and cordials 50
Syrups 50
2.2.9 Preserves 53
Jams, jellies and marmalades 53
Pastes and purees 53
Fruit cheeses 53
Additional processing notes 54
Batch preparation 54
Boiling 55
Filling 56
2.2.10 Wines, vinegars and spirits 56
Additional processing notes 62
2.3 Conducting a feasibility study 63
2.3.1 Introduction 63
2.3.2 Market analysis 64
Product quality survey 65
Survey of market size and value 67
Market share and competition 71
2.3.3 Technical feasibility 75
Production planning 77
Weights of raw materials and ingredients required 79
Equipment required 82
Packaging 83
Staffing levels 83
2.3.4 Financial feasibility 87
Start-up costs 87
Operating costs 88
Income and profit 90
Financial planning 92
Preparing a business plan 93
2.4 Legal aspects 95

Page

2.4.1 Registration of the enterprise 95
2.4.2 Food-related laws 97
Food composition 98
Food labelling 100
Hygiene and sanitation 101
Weights and measures 102

2.5 Establishing production facilities 103
2.5.1 Introduction 103
2.5.2 The site 103
2.5.3 Design and construction of the building 103
Roof and ceilings 104
Walls 104
Windows and doors 105
Floors 105
Lighting and power 106
Water supply and sanitation 107
Layout of equipment and facilities 109
2.5.4 Equipment 109
Dried products 113
Boiled, concentrated and pasteurised products 115
Fermented and distilled products 118
Packaging, filling and sealing equipment 119
2.5.5 Packaging materials 121

2.6 Contracts with suppliers and retailers 123
2.6.1 Securing raw materials 123
2.6.2 Agreements with retailers and other sellers 126

2.7 Managing production and quality assurance 129
2.7.1 Managing production 129
Production planning 129
Scheduling inputs 130
Maintenance 131
Staff management 132
Health and safety 134
2.7.2 Managing quality assurance 135
Safety of products 135
Product quality 139
Raw materials and ingredients 139
Processing 143
Packaging, storage and distribution 147
Hygiene and sanitation 150

2.8 Marketing 152
2.8.1 Identification of markets 152
Market segments 152

Page

		Page
	Distribution and promotion	154
	2.8.2 Developing a marketing strategy	155
	2.8.3 Packaging and brand image	158
2.9	**Record keeping**	161
	2.9.1 Financial and sales records	162
	2.9.2 Production records	166

Appendices

Appendix I	Basic rules for hygiene, sanitation and safety	169
Appendix II	Checklist of entrepreneurial characteristics	171
Appendix III	Feasibility study checklist	173
Appendix IV	Institutes that support small scale food processing	175

Bibliography 181

Index 191

I General introduction

Food processing as a scientific and technological activity covers a broader area than food preparation and cooking. It involves the application of scientific principles to slow down the natural processes of food decay caused by micro-organisms, enzymes in the food or environmental factors such as heat, moisture and sunlight - and so preserve the food. Much of this knowledge is known traditionally and put into practice by experience and information handed down through the generations. In most developing countries food processing is also a method of generating employment and family incomes. Under these circumstances, producers must compete with others in the same country and with imported products. With some important exceptions, traditional processing methods produce foods that are usually inadequate to compete with the 'newer' products. This is particularly important with respect to packaging and presentation of the processed foods.

One purpose of this book is to document improved ways of processing fruits and vegetables at a scale that is appropriate for families and other groups of people to earn a daily income. This also involves methods of business planning, work organization, marketing and quality assurance that are likely to be unfamiliar to traditional processors, but which are essential to ensure successful and profitable production.

There are also important reasons for improving food processing that is done in the home to meet family needs and the book is therefore divided into two Parts: first a description of some of the problems and issues that face people involved in home processing, together with some suggested solutions; and secondly a more detailed description of the various aspects that are involved in setting up a small food processing enterprise. In general terms, the benefits of small scale fruit and vegetable processing for people in developing countries can be broadly stated as follows:

- raw materials are readily available (often in surplus)
- most technologies are available, accessible and affordable at scales that are suitable for small operations
- equipment can often be manufactured locally, creating additional employment
- the products, if chosen correctly, have a widespread demand
- compared to some other technologies, small-scale fruit and vegetable processing is particularly suitable for women
- most processes have few environmental impacts.

However, the selection of suitable products for small scale manufacture and the processes chosen to make them, require very careful consideration. It is not sufficient to assume, as many 'advisers' do, that simply because there is a surplus of a raw material each year, a viable fruit and vegetable processing venture can be created to use up the excess. There must be a demand for the processed food which is clearly identified before a business is set up. Otherwise the most likely result is to produce a processed commodity that no-one wants to buy with substantial financial losses to those involved.

There are numerous examples of misguided development projects that have resulted from a desire to prevent piles of rotting fruit from accumulating each season by processing them. Instead, the result has been shelves of more expensive, rotting processed foods that no-one wants to eat.

In general the types of products that are suitable for small scale production are those for which there is a high demand, and a higher value can be added by processing. Typically fruits and vegetables have a low price when in their raw state, but can be processed into a range of snackfoods, dried foods, juices, pickles, chutneys etc., which have a considerably higher value.

The high added-value means that the amount of food that must be processed to earn a reasonable income is relatively small. Hence the size and type of equipment required to operate at this scale can kept to levels that are affordable to most aspiring entrepreneurs.

A further consideration that applies to both home processing and commercial production is the safety of the fruit and vegetable products that are produced. Although there are many similarities between fruit and vegetables, it is important to note the following difference: most fruits are acidic and this acidity controls the type of micro-organisms that are able to grow in the processed products. Food poisoning bacteria cannot grow in the acidic conditions, but moulds and yeasts are able to grow and produce obvious visible spoilage which stops consumers from eating the food. Even if a contaminated fruit product is eaten, yeasts and moulds rarely cause severe food poisoning.

Vegetables however, are generally less acidic than fruits and a wider range of micro-organisms are able to grow in vegetable products, including food poisoning bacteria. This is particularly dangerous when certain types of bacteria release poisons (or 'toxins') into the food, but do not produce obvious signs of spoilage. Consumers may therefore be unaware of the contaminating bacteria and eat the poisoned food. Careful control of processing conditions and attention to hygiene are therefore essential when processing less acidic vegetable products.

Dried fruits and vegetables do not allow micro-organisms to grow, provided that drying is carried out correctly and the food is kept dry. However, if the food is heavily contaminated before drying, or if it is allowed to get damp during storage, these micro-organisms can grow again before the food is eaten. A summary of the effects of different processing methods on micro-organisms is shown in Table 1.

In addition to the actions of micro-organisms, naturally occurring enzymes rapidly change the colour, flavour and texture of fruits and vegetables after harvest. In general, these changes, due to both micro-organisms and enzymes, produce a characteristically short shelf life and rapid processing after harvest is therefore necessary.

Details of the points in a process where control is needed to ensure food safety or to maintain food quality are given in the descriptions of each production process in Part 2 and a summary of good food handling practices is given in Appendix I.

In many countries, vegetables and fruits are among the most accessible raw materials for processing. Traditionally, cultivation of vegetables in small garden plots has been common in most tropical regions and the planting of fruit trees around the house or compound has provided shade and wood as well as fruit.

These activities are being expanded and promoted in a large number of both home gardening programmes to improve household nutrition and in orchard planting and regeneration programmes designed to address environmental concerns such as soil erosion and also to improve fruit quality and varieties.

These programmes, together with an increasing awareness of the value of processing for improved food security and income generation, has resulted in fruit and vegetable processing being seen by many development agencies and government institutions as an important method to improve the livelihoods of both rural and urban populations.

However, to be successful these initiatives should from the outset have clearly defined objectives concerning the expected benefits of processing. These may include direct benefits to participating groups or individuals, such as:

- Improved storage of fresh produce without excessive losses
- improved nutritional status through consumption of fruits and vegetables for a larger part of the year
- increased income for sale of processed fruits and vegetables
- preservation of seasonal gluts which would otherwise be wasted.

There may also be improved backward linkages to farmers and suppliers (especially in integrated development programmes) as a result of increased demand for raw materials by processors. This may:

- improve the amount, quality and varieties of fruits and vegetables that are grown

Processing method	Principle of preservation and effect on micro-organisms
Preparation methods: Washing Sorting Slicing, chopping, mixing, mincing, etc.	No preservative action. No destruction of micro-organisms (some processes such as chopping or mixing may actually promote growth of micro-organisms by making the food more available for them to grow on)
Heating: Blanching Boiling Pasteurizing Canning/bottling Concentrating* Extruding* Baking/roasting Frying*	Heat kills all types of micro-organisms and destroys the ability of naturally occurring enzymes to act. Mild heating (such as blanching and pasteurization) kills some but not all micro-organisms. Different types of micro-organisms have different degrees of resistance to heat. Other types of more severe heating, such as canning, kill most micro-organisms.
Removing heat: Freezing* Chilling	These processes slow down the rate at which micro-organisms can grow and enzymes can act, but processing does not destroy them. Freezing also turns water to ice so that it is not available for micro-organisms to use. Freezing therefore has a greater preservative effect than chilling.
Removing water: Drying Distilling Other processes*	Water is needed for micro-organisms to grow and for enzymes to act and drying therefore prevents this. However, many types of micro-organisms are not killed by lack of water and they can grow again when food is dehydrated. Water is also removed by other processes (*) which has a combined effect in preserving the food.
Chemical: Fermenting to produce acids or alcohol Adding acids, sugar*, salt* or preservatives.	Micro-organisms are killed by high concentrations of alcohol, salt or sugar. Different types of micro-organisms have different levels of resistance to these chemicals and in general moulds and yeasts are able to grow whereas bacteria cannot.

* Water is removed or made unavailable, which contributes to the preservative effect.

Table 1. - Effects of different types of food processing on micro-organisms

(Note: the use of low temperature processing is not considered in this book due to the high capital costs of equipment and the need for refrigerated distribution and storage, which is both expensive and not widely available in most developing countries).

- improve the rural environment as a result of planting fruits and vegetables. Such improvements may include soil regeneration and rainwater retention due to orchard or vegetable planting

- improve incomes to farmers from higher sales of raw materials to processors, or secure supply contracts that reduce farmers' dependency on seasonal price fluctuations
- increase demand for equipment, packaging materials and ingredients which stimulates development of associated supplier industries and creates strategic alliances between different manufacturing sectors.

There may also be benefits from improved forward linkages between processors and retailers/customers including:

- lower cost and increased availability of locally processed foods
- improved quality through greater competition between producers
- increased diversification in the varieties of processed food that are available in the diet
- improved relationships and trust between processors, retailers and customers.

When sufficient numbers of new processors are operating there may also be an effect on the national economy due to reductions in the use of foreign exchange by substitution of imported ingredients or finished products.

When promoting food processing, each programme should therefore have a clearly defined set of objectives. These can be broadly characterised as:

a) those that are likely to result in improved nutritional status or better food security; and

b) those that are likely to improve individual incomes or national economies.

Different approaches are needed when promoting processing to increase food security or nutritional status, compared to those that are used to establish or support small processing enterprises for income generation. The different objectives of a programme should therefore be clearly identified at the outset and appropriate methodologies developed to address each one. Although details of such methodologies are beyond the scope of this book, a summary of issues that should be addressed to meet differing objectives is shown in Table 2.

In this book, aspects of improved nutrition and food security are described in Part 1 and those concerned with processing for income generation are included in Part 2.

Issues that affect food security and nutritional improvement programmes	Issues that affect enterprise development programmes
• Nutrition education • Health and hygiene training • Development of groups • Improved communication • Confidence building • Agricultural skills training • Credit support systems • Development of producer cooperatives or other farmer associations • Improvements to equipment and tools • Improved transport and infrastructure • Seed banks and other sources of agricultural inputs	• Market awareness and consumer preferences • Marketing strategies, promotion and packaging • Methods of financial control • Developing trusting relationships with suppliers and retailers • Quality assurance • Hygiene and sanitation for production • Food legislation • Taxation and business legislation • Training in management and business planning • Finance and credit suppliers • Staff training in production technologies • Suppliers of specialist equipment and ingredients

Table 2. - Issues to consider when addressing different objectives of food processing

PART 1

PROCESSING FOR HOME CONSUMPTION

1.1 Introduction

The majority of fruits and vegetables grow in short seasons and one main reason for processing is therefore to secure a supply of these foods when they are not available in a fresh state. This is both because some (tomatoes, onions, spices etc.) are used in daily cooking in different societies and because regular consumption of fruits and vegetables is necessary to maintain health.

In different societies throughout the world, the consumption of fruits and vegetables has different levels of significance in the diet. In some areas, such as the cold, infertile conditions of Central Asian steppes, mountainous regions of Latin America and arid deserts of Africa and Asia, fruit and vegetable cultivation is difficult and has a short annual growing season. These products do not therefore play an important role in local diets and there is greater reliance on animal products. However, the requirement to eat fruit and vegetable products to maintain health means that in almost all of these societies, small amounts of leafy vegetables and collected fruits are preserved for winter or dry seasons and as a result of their shortage, they have a high value.

By contrast, in almost every country of the humid tropics and sub-tropical regions, the soils and climate are suitable for fruit and vegetable cultivation and there is an historic reliance on these products as an important part of the normal daily diet. Indeed in some areas this may be reinforced by religious beliefs, such as Hinduism, which requires followers to be vegetarians. In many communities throughout the world, vegetables are eaten daily as an accompaniment to a main cereal or root crop meal, with or without meat and fish. Fruits are eaten fresh when in season as appetisers, snacks, juices and desserts.

There are a large number of traditional processed fruit and vegetable products that are traditionally made in the home. Examples from Asia include fruit 'leathers' (dried sheets of fruit paste that have an appearance of thin leather), fruit pastes, pickles and chutneys. In Africa and Latin America, a wide range of dried chips, dried and powdered leaves, fruit beers and other fermented fruit and vegetable products are traditionally produced. Each of these products is stored for future use and is often not intended for sale, although no clear distinction is possible. Some products that are processed for home consumption may also be sold to neighbours or at village markets when additional income is needed. Problems that may be found when developing this type of income generating activity are described at the end of this Section.

In general terms because of low incomes, home processing requires simple, cheap equipment and will usually be done without specialised machines, using existing kitchen utensils and simple pieces that can be made from local materials at almost no cost. Processing methods are also inexpensive but in some cases are complex and sophisticated, involving a sequence of process stages that require considerable experience and expertise to perform correctly.

Some of the main reasons for home processing are to maintain a secure supply of food for seasons of shortage (to provide food security) and to maintain health through an adequate supply of nutrients in the diet. These aspects are described in more detail below.

1.2 Food Security, Nutrition and Health

Fruits and vegetables provide an abundant, cheap source of fibre and several vitamins and minerals (Table 3). In general, they have the highest nutritional value when eaten fresh, although an exception may be fermented foods, in which the process of fermentation can increase the content of B-vitamins (Table 5).

However, fresh fruits and vegetables have a relatively short life before they begin to decay and if they are to be stored for an extended period, it is necessary to process them by one or more of the various methods that are available. The extent to which preservation is required depends in part on the type of plant material that is being considered. For example, fast-growing parts such as shoots have a short life before rotting, whereas bulbs and tubers can be stored without preservation for much longer at room temperature (25°C) (Table 4).

The extent to which nutrients are lost during processing varies according to both the type of fruit or vegetable and the process that is used. A summary of losses of selected nutrients is shown in Table 5.

The purposes of processing fruits and vegetables in the home may therefore be summarised as follows:

- preservation of seasonal gluts for survival in times of shortage
- provision of cooking ingredients throughout the year
- increase in convenience by having part-prepared foods available when required.
- availability of different foods throughout the year to provide variety and greater nutritional balance in the diet.

Food	Fibre (g/100g)	Carotene (µg/100g)	Thiamine (mg/100g)	Ascorbic acid (Vitamin C) (mg/100g)	Calcium (mg/100g)	Iron (mg/100g)
Soya flour	11.9	-	0.75	0	210	6.9
Wholemeal bread	8.5	0	0.26	0	23	2.5
Boiled rice	2.4	0	0.01	0	1	0.2
Doughnut	-	0	-	-	70	1.9
Whole goats' milk	0	22	0.04	1.5	130	0.04
Yoghurt	0	5	0.05	0.4	180	0.09
Eggs (raw)	0	tr	0.09	0	52	2.0
Beef (raw)	0	tr	0.05	0	7	1.9
Chicken (raw)	0	tr	0.10	0	10	0.7
Green beans (raw)	2.9	400	0.05	20	27	0.8
Green cabbage (boiled)	2.2	500	0.03	25	30	0.5
Carrot (raw)	2.9	12,000	0.06	6	48	0.6
Okra (raw)	1.3	90	0.10	25	70	1.0
Plantain (raw)	5.8	60	0.05	20	7	0.5
Tomato (raw)	1.5	600	0.06	20	13	0.4
Apricot (raw)	2.1	1,500	0.04	7	17	0.4
Banana (raw)	3.4	200	0.04	10	7	0.4
Mango (raw)	1.5	1,200	0.03	30	10	0.5
Orange (raw)	2.0	38	0.08	38	31	0.3
Pineapple (raw)	1.2	60	0.08	25	12	0.4

tr: = trace amount

Table 3. - Dietary fibre, selected vitamins and minerals in fruits and vegetables compared with some other foods.
(Data adapted from: *The Composition of Foods,* by Paul and Southgate)

Part of plant	Time before spoilage starts (days)
Fast growing shoots	1-2
Leaves	2-3
Stems	5-50
Fruits	10-30
Roots and tubers	15-50
Bulbs	30-100

Table 4. - Rate of spoilage of different parts of fruit and vegetable plants

When the objective of a programme is to improve health and nutritional status by increasing the effectiveness of food processing, it is usually necessary to combine education and training with practical improvements to facilities and equipment. In order for people to learn from each others' experience and to make such interventions cost effective, it is usually necessary for participants to organise themselves into groups. This in itself can be a long and difficult process in some communities, but where groups such as mothers' clubs or farmers' societies already exist, they can be approached to act as a focus for a new programme. In some instances, successful food processing developments for improved nutrition have also been linked to health centres.

Any improvements to facilities and equipment need to be low cost and easily maintained if they are to be used in a sustained way. Whereas the capital cost of a new building or item of processing equipment can be donated, the facility will cease to operate if both funds and responsibility for maintenance are not established at the outset. Many attempts have been made to share the costs of processing by establishing various forms of producer associations or village co-operatives, with varying degrees of success.

One important factor is the degree of trust and willingness to work together that exists in a community. If there is no history of a communal approach to problem solving or implementing a new development, it can take many years to introduce these ideas. However, there are many examples of successful village associations that have been operating for several years and have supported on-going education and training programmes in health and nutrition, together with the establishment of community systems and facilities for crop processing.

1.2.1 Improvements to home processing and storage

In most cases, simple methods of processing and storage that are based on traditional knowledge have the greatest chance of being successfully introduced. Improvements to existing practices are more likely to succeed than completely new technologies, not only because they are more familiar to people, but also because traditional methods are suited to local environmental conditions, whereas new technologies may not be. A good example of such an approach is improved drying and storage.

Drying

Drying is one of the most accessible and hence the most widespread processing technology. When combined with improved food stores, it can lead to significant improvements in food security in most regions (exceptions being some areas that have a climate in which drying is difficult for the majority of the year without the use of fuel - see Section 2.2.3). Important examples of dried fruits and vegetables are okra, cabbage, spinach, mango slices or leathers, garlic and other flavourings, all of which are stored for use in daily cooking.

	Loss (%)						
Process/product	Carotene	Thiamine	Riboflavin	Niacin	Vitamin C	Panto-thenic acid	Vitamin B_6
<u>Fermentation</u>:							
Grapes/wine			66	66			
Cabbage/pickled			43	33	73	57	19
Soybean/Tempeh		41	(+817)	(+488)			(+437)
Soybean/soysauce		(+400)	(+616)	(+667)			
<u>Blanching</u>:							
Peas					25		
Broccoli					30		
Green beans					20		
<u>Canning/bottling</u>:							
Carrots	0-9	67	38-60	32	75	54	80
Peas	0-30	75	47	71	67	80	69
Tomatoes	0	17	25	0	26	30	10
peaches	65	49	39	39	56	71	21
Pears	-	45	45	0	73	69	18
Pineapple	25	7	30	0	57	12	-
<u>Drying</u>:							
fruits (average for apple, apricot, peach and grape)	6	55	0	10	56	-	-
Vegetables (average for peas, maize, cabbage and beans)	5	<10	<10	-	-	-	-

Table 5. - Summary of nutrient losses in different types of food processing
(Adapted from: *Food Processing Technology* by Fellows)

Sun drying is traditionally practised because there is negligible cost in processing and the work of spreading and turning the crop as it dries can be shared among family members, including young children. Some of the reasons why the introduction of solar dryers into village communities has not been widely successful (Section 2.2.3) concern these factors of cost and workload. Even when constructed from locally available materials, the simplest dryer still has a higher cost than sun drying. Additionally, the work involved in loading and carrying trays of fruits and vegetables and removing them regularly to turn and mix the products, is not suitable for children and therefore places an additional workload on the parents, usually the mothers. Furthermore, the possibility of improved quality products and shorter drying times using solar dryers may not be important factors to families who may be satisfied with the existing quality of dried foods and the time and effort required to dry them.

Under such circumstances, a focus on improvements to sun drying is likely to yield greater results. Examples include raising crops off the ground on woven matting or wooden drying frames to reduce contamination by dust, crawling insects or rodents; covering the crop with mosquito netting to reduce damage and contamination from birds and flying insects; cutting the fruits and vegetables into smaller pieces so that they dry faster and blanching vegetables to retain a better colour. Although it is beyond the scope of this book, it may be possible in some programmes to link drying improvements to agricultural developments and the introduction of crop varieties that are better suited to drying than traditional types.

Concentration by boiling

Relatively few fruits and vegetables are preserved in the home by concentration, mainly because of the additional fuel-wood consumption, but important foods in some societies are tomato paste, extracts from wild plants such as baobab fruit, that are used as flavourings, chutneys and syrups made from fruits or saps. The main requirement to improve processing of these products is to control the rate of heating to prevent localised burning of the product, particularly when it has become thickened towards the end of boiling.

Fermentation

Production of fruit wines and beers is possibly the second most common form of home processing after drying, in countries where alcohol consumption is not prohibited by religious beliefs. A wide range of fruits is used, particularly pineapple, melon and pawpaw. In the example below from East Africa, banana is used. It is difficult to produce a clear juice from bananas but the following description by the processor indicates the complexity and degree of skill required to produce a satisfactory product.

'Bananas are harvested when they are not quite ripe and not over-ripe. They are stored in a pit covered with banana leaves for up to 4 days to ripen. If necessary a fire is lit in a small hole leading into the pit to warm the bananas and hasten ripening. The bananas are peeled and placed into a large wooden trough. They are then crushed to a pulp by walking on the fruit. Two grasses, one of which is young elephant grass, are cut to approximately 15cm long and carefully mixed in the correct proportions. Thin layers are periodically added to the surface as pulping proceeds and the grass becomes mixed into the pulp. This is to finely divide the pulp into small particles and thus release the juice. Only these two grasses are used as they do not add flavour to the juice in the way that other species do'.

Figure 1. - Traditional pulping of bananas to make juice for fermentation
(Courtesy of Midway Technology)

'After approximately 30-45 minutes pulping, the mixed grass and pulp is piled on a coarse filter frame, made from wooden poles lashed together over the trough, and juice is pressed out by standing on the pulp. The juice is then collected using a calabash into which a hole is cut in one side and a hollow handle is fitted to the top. Juice is poured through the hollow handle into a second calabash, which is shaped as a funnel and lined with grass to act as a fine filter. The juice, which is cloudy but without particles, is collected in a clean container. It is tasted at this stage and if necessary, adjusted to the correct sugar content by adding water.

The standardised juice is then returned to a clean wooden trough and sorghum is added in the ratio of 2kg per 20 litres of juice. The trough is covered with banana leaves and also with the residue from pulping to act as insulation and maintain the temperature of fermentation at ambient (22-26°C) for two days. After fermentation, the liquor is strained through the calabash/grass filter into clean containers or bottles and is then ready for drinking. There are no further clarification or pasteurisation stages and the beer is preserved by the alcohol content (3-5%). As a result it has a shelf life of up to four days without refrigeration. The process could be improved and a longer shelf life obtained, if the product was clarified by filtration or sedimentation and pasteurised in the bottles.'

Home preservation of fruits and vegetables by fermentation to produce pickles is especially common in South and Central Asia, but less so in other regions of the world. Examples include pickles from lime, mango, mixed fruit and a range of vegetables, including different types of cabbage. Each local district may have a slightly different set of ingredients or variations in the process and in India for example, there are estimated to be over 700 different types of fermented pickle.

In all cases the prepared fruit or vegetable is mixed with spices and held in a closed container such as glazed pottery, while a natural sequence of bacteria ferment sugars in the raw materials to produce lactic acid. Provided that the raw materials are not heavily infected with moulds and that insects, which can carry yeasts or moulds on their bodies, are kept out of the container, the build up of lactic acid is sufficient to prevent spoilage and the product can be kept for several months while it is being used.

In the following example from Central Asia, pickled cabbage is produced: a 2cm layer of salt is placed into the bottom of a wide-necked pottery drum or jar. A 5cm layer of shredded cabbage is placed onto the salt and the jar is then filled using alternating layers of salt and cabbage. It is then sealed with a wooden or cloth bung and allowed to ferment for several weeks. As the water is drawn out of the cabbage, it forms a pickling brine. If the product is not to be used straight away, the cabbage is weighted down with a clean stone to keep it beneath the brine and to prevent it drying out and spoiling. Little improvement can be made to these processes without substantially increasing capital and/or operating costs. They are well suited to local tastes and environmental conditions.

Pickling

There is also a wide range of unfermented pickles produced in Asia. Two examples below describe the production of lime pickle and mixed pickle from South Asia.

Well ripened, but not over-ripe limes are washed, cut into four (but the pieces are not separated), and dipped in a concentrated solution of salt. They are then dried in the sun for several days. Sometimes salt crystals are sprinkled onto the lime while it is drying. Drying is continued until the skin develops a brown colour and the pieces can be broken using the fingers. The product is then packed along with the expelled juice in pottery vessels. The salt removes water from the pieces due to osmosis and some sugar is also removed. As a result, salt-tolerant micro-organisms begin to grow while the product is being dried. These bacteria produce some acid, but mostly generate the characteristic flavour of the product, which has a sour and salty taste and can be kept for several months.

The main raw materials for mixed pickle are green chillies, onion, papaya, and spices. The papaya and onion are peeled and cut into slices. The spices are also washed and then ground. All ingredients are placed in a clean pot and mixed by hand while adding vinegar. The product can be kept in the sealed pot for several months, during which time there is a degree of fermentation which improves the flavour, but it does not contribute to the preservation. Other formulations of mixed pickles use raw materials such as dates, pineapple or mango,

Storage

In many cases, the most serious problems of food security are connected with inadequate food storage practices. For example, dried foods are susceptible to spoilage by mould or to attack by insects and rodents, particularly if they are inadequately dried. Losses during storage can be considerable and when these are cereal losses, they can be life threatening. Traditional storage practices for fruits and vegetables vary widely in different parts of the world, but all use locally available materials, usually at little or no cost. It is not possible in a book of this size to detail these differences, but the general principles of storage are described below, followed by some examples of good practice that may assist in improving storage practices and facilities in other places.

As described above, the main types of processed fruits and vegetables that are stored in the home are dried foods, followed by fermented and pickled foods and concentrated syrups or pastes. The latter groups are each moist foods but they are preserved by raised concentrations of lactic acid, salt or sugar. All that is required for adequate storage is therefore an impervious pot to contain the contents and an airtight and insect or rodent proof seal to prevent contamination or theft. Glazed pottery or glass vessels, with cork or wooden stoppers, sealed with resin or wax are entirely adequate for the expected shelf life of these products. They should be stored off the floor in a cool dark place, away from sunlight and dampness.

Dried fruits and vegetables need protection from insects, animals and from uptake of moisture. Those foods that are required to remain unbroken also need protection from crushing and those that contain high levels of carotene, such as green leafy vegetables or carrots, also require protection against sunlight. Because of their bulkiness, pottery vessels may be too expensive to store an adequate supply and many families resort to hanging dried fruits and vegetables from rafters or over cooking areas, where they are sure to remain dry. This deters most animals but does not provide adequate protection against insects. Improved storage can be achieved by constructing a simple mesh basket from wooden sticks and mosquito netting, which can be hung from the rafters in a cool, dry place, away from sunlight.

Dried foods should not be stored on the floor. If storage on tables or raised platforms in the house is the only option, the supporting legs should rest in small containers of oil or kerosene to deter crawling insects. Storage containers should be protected by insect proof mesh. Ideally, re-used cans, having press-on lids or similar insect- and rodent-proof containers are used, once they have been thoroughly cleaned and dried.

1.2.2 Home processing to earn extra family income

There are very large numbers of people throughout the developing world who process foods in their homes and sell them to neighbours or in local markets, in order to generate a little extra money to supplement family incomes. People in this informal sector are characterised by having little capital to invest in equipment and they are forced to buy inputs such as raw materials, ingredients and packaging in small quantities from retailers, because of a lack of working capital or an uncontrolled cashflow (see Section 2.3.4). This increases the unit costs of their products and may make them un-competitive on price with those from the formal sector, or even in some cases with subsidised imports. They have poor access to credit because they have little or no collateral and credit providers cannot justify the high administrative costs of servicing a large number of small loans.

Figure 2. - Informal sector processing of fruit

It is also frequently the case that informal processors have little experience of quality control and so their products do not compete well on quality, safety or appearance with those from the formal sector. Because they only process fruits or vegetables until they have earned sufficient money for their immediate needs and they then cease production until the next financial need, informal enterprises are unable to build up trusting relationships with suppliers or retailers and as a result, future orders may be more difficult to obtain.

Thus a complex mix of factors need to be addressed when a programme is designed to assist informal, home-based processors. Provision of accessible credit is one component, but by itself is insufficient to generate success. Technical training in production routines, hygiene and quality assurance is also required and most importantly, an improved ability to identify suitable markets and to develop skills in selling the products is needed. The different objectives of a programme of assistance, described in Section I, should be clearly identified when initiating work of this kind.

There are a number of development programmes that are achieving some success in improving the incomes to processors in the informal sector and some of the organisations listed in Appendix IV have experience in this area. It is not usually possible for an individual farmer or rural family to adequately learn all of the skills needed to process and sell fruit and vegetable products in competition with large scale producers. But the formation of groups of people that are able work well together with mutual trust among the members has in some cases been shown to overcome these problems. People learn from each other, are empowered by the group, are able to access credit because of the larger needs of the group and in some cases, are able to achieve savings by bulk purchases of inputs.

The division of labour within the group also has social benefits not only in providing a more flexible working environment to take account of household responsibilities but also in developing, encouraging and rewarding an individual's competence and skills in their area of work. This can have dramatic effects on the confidence and determination of individuals to succeed. This approach also removes the need for each member of the group to have expertise in all areas of operating a processing enterprise (for example checking quality, negotiations with suppliers or customers, keeping records, producing financial data, marketing and selling etc.), and allows each member to develop their own specific contribution to the success of the group, based on their individual skills and abilities. There can be various structures for such groups, including producer co-operatives, in which each member has an equal share, groups based on religious beliefs, extended families or those formed around a different existing activity, such as a youth club, a mothers' union or a village development society.

A further development of group activity in the informal sector is the division of responsibility for different parts of the food supply chain. A number of groups can reach agreement to collaborate and share the profits of the joint venture. In some successful approaches, a farmers' group is responsible for supplying a local group of processors with an agreed amount of raw materials of specified quality. This removes the need for farmers to develop skills in marketing their crops and they can concentrate on their farming abilities. The processors may buy from a number of farmers in the local area, thus reducing transport costs for the raw materials. The processors concentrate on producing high quality products at the lowest cost and sell them to a marketing group. The marketing group may buy from a number of processors and concentrate only on finding the highest paying and most reliable markets for the products.

A development of this system may involve part-processing of fruits or vegetables on the farm to reduce transport costs and to add value to the farmers' produce. For example, fruits can be peeled and part-dried or crystallised on the farm and then transported to a local processor in returnable containers. There are some environmental advantages in doing this because not only are fuel costs for transportation reduced, but also localised pollution at the central processing unit is removed and fruit and vegetable wastes are available on the farm for green manure. In this system, the processor standardises the quality of each batch by blending or sorting and then packages each farmers' products under a single brand name. This gives both benefits of lower costs through bulk purchases of packaging and the ability to supply larger amounts to meet demands of retailers who would otherwise not consider purchasing.

There is thus a spectrum of processing enterprises from a single person working intermittently from home, through various forms of single and group-based informal enterprises to more formal structures. In Part 2 of this book, the detailed aspects of operating formal food processing businesses are described.

PART 2

PROCESSING FOR SALE

2.1 Introduction

Interest in small scale food processing as a means of enterprise creation in developing countries has increased substantially in recent years, as a result of a number of factors. These include:

- greater promotion of private sector micro- and small-scale enterprises by national and international development agencies, to help increase incomes and employment opportunities in both rural and urban areas
- development of tastes for non-traditional foods, especially in major urban centres
- a desire by national governments to both increase export opportunities and foreign exchange earnings and to develop national manufacturing capacities
- increased food surpluses in some countries that require preservation and processing as a consequence of successful agricultural interventions or development programmes
- an increasing pride in locally produced products that can effectively compete with imported goods.

However, successful fruit and vegetable processing enterprises require more than the skills that are needed for home processing, described in Part 1. Whereas in home or village processing, the consumers and processors know each other and can give feedback on the quality or perceived value of a processed food, in formal food processing businesses the producers usually do not know who eats their food or what their customers think of the products. Similarly the customers have no direct link to the individuals who have produced their food and there are thus very limited opportunities for feedback between processors and consumers. As a result, processors have to learn new skills in developing attractive packaging and in marketing and selling (Section 2.8.3) if they are to successfully find customers.

Additionally, consumers expect to find the same food in every pack, having a uniform quality, every time that they buy a product. Producers must therefore control their process to produce uniformity and consistency in their products, and to do this they must learn new skills in quality assurance. There are also other problems that are associated with operating any enterprise that need to be addressed by aspiring entrepreneurs: for example legal aspects such as registration of the business, payment of taxes, employment conditions and business planning. These issues are discussed further in Sections 2.4-2.8 of this Part of the book.

For those who set out to earn their main income from fruit and vegetable processing. there are special problems which make this type of business different from most others:

- many raw materials are highly perishable and will spoil quickly after harvest unless they are processed quickly
- many fruits and vegetables are also highly seasonal which means that the business can only operate for part of the year, unless crops are either part-processed for intermediate storage, or a succession of crops is processed throughout the year (Figure 3)
- seasonality also affects the cashflow in an enterprise, as most raw materials have to be bought during a relatively short harvest period
- yields of fruit and vegetables are subject to considerable variation according to the weather, especially the rainfall patterns, and plant diseases. This can result in an unpredictable supply and large variations in prices for raw materials, which makes business planning more difficult (Section 2.7.1)
- some processed foods also have a seasonal demand (for festivals, ceremonies etc.) which further complicates planning and cashflow of a business
- even after processing, some fruit and vegetable products have a limited shelf-life and distribution and sales methods must therefore be organized so that customers receive products in the required amounts before they spoil
- processing must be done to high standards of hygiene and production control to avoid the risk of harming or even killing customers by allowing contamination of the products by foreign materials or the growth of food poisoning micro-organisms.

Fruit and vegetable processing as a business is therefore a more complex activity than processing at home. Although it is true that most people can be trained to make high quality products, it is misleading to say that a person who is able to make a good product at home can then become a successful entrepreneur. Many (possibly most) small scale food processors start by working from home, but their ultimate success is dependent on a number of factors, including their planning and business skills, their creative flair to produce products that have a demand and are different from those of competitors and their determination to succeed.

Characteristically, successful entrepreneurs see food processing as their main source of income. They may take out a loan to buy specialized equipment or secure working capital and they develop business and marketing skills to expand and diversify their enterprise. A questionnaire about the entrepreneurial characteristics that are required to successfully develop a small fruit and vegetable processing business is given in Appendix II.

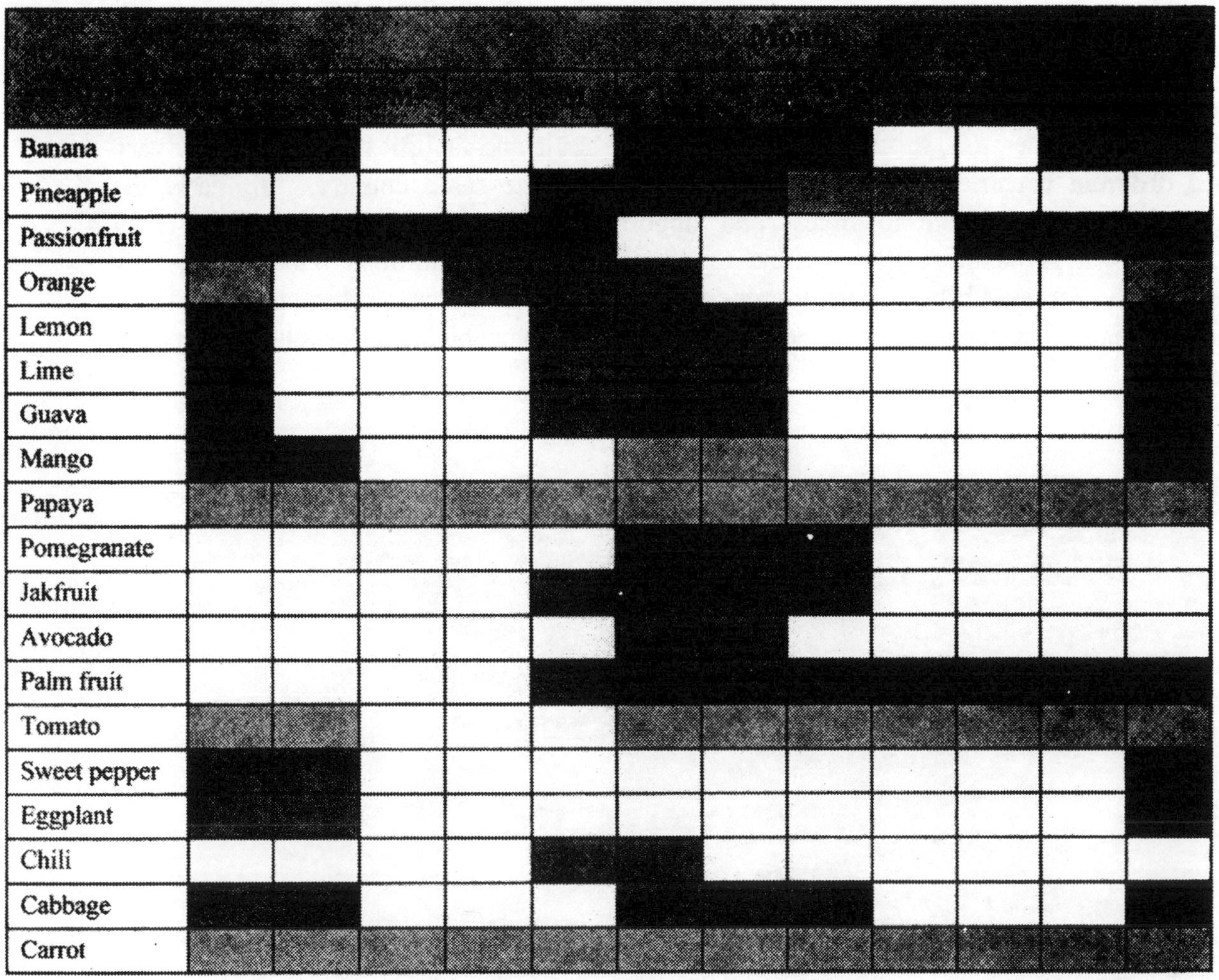

Figure 3. - Example of a seasonality chart for fruits and vegetables

Main harvest season

Crop available

Rainfall in March to May and August to November.

There are different definitions of micro- and small-scale enterprises and in this book, the following are used:

> A micro-enterprise is one in which the owner works each day to manage the business and produce food. The number of workers is less than 10, the total investment does not exceed $20,000 and annual sales are less than $12,000.
>
> A small scale enterprise is one in which the owners may work at the processing unit or employ a manager. The number of workers is less than 20, the total investment does not exceed $50,000 and annual sales are less than $25,000 (average for African countries).

The different skills required by producers to successfully operate a small enterprise and earn a family income are described in this Part of the book. The availability of services (electricity, clean water, gas, servicing and maintenance skills and facilities etc.) obviously vary between different countries and even between regions of the same country. Similarly the amount of money available to invest and opportunities for obtaining credit also vary considerably. It is therefore not possible to describe precise conditions under which any business can be successful, but where appropriate, guidelines are given on the types of activity and investments that have made other small fruit and vegetable processing enterprises successful.

Figure 4. - Products from a successful fruit processor
(Courtesy of Midway Technology)

The criteria that need to be taken into account when deciding which technologies to use are complex and inter-related but are likely to include the following:

- technical effectiveness (will the equipment do the job required at the indicated scale of production
- relative costs for both purchase and maintenance of equipment and any ancillary services required
- operating costs and overall financial profitability
- health and safety measures in the process
- conformity with existing administrative or production conditions
- training and skill levels required for operation and maintenance

- environmental impact, such as pollution of air or local waterways.

However, these factors are not simply a checklist and each will have a different weighting in different circumstances. There can be no simple solution to the difficult task of assessing all factors in a particular situation and making the 'best-fit' from the available solutions.

In every proposed development, the general principle when starting a business, having decided which product to make, is as follows: conduct market research and produce a draft feasibility study. If the business appears to be profitable, produce a business plan, obtain credit where necessary and register the business with taxation and local government authorities. Then establish production facilities, develop contracts or agreements with suppliers and customers and manage the routine production of the products. Continually review and update business plans to take account of the actions of competitors or changing markets.

The Sections below follow this sequence, starting with a description of product characteristics and production methods for a range of fruit and vegetable products, to assist potential entrepreneurs in deciding which products to make. These are summarized in Figure 5.

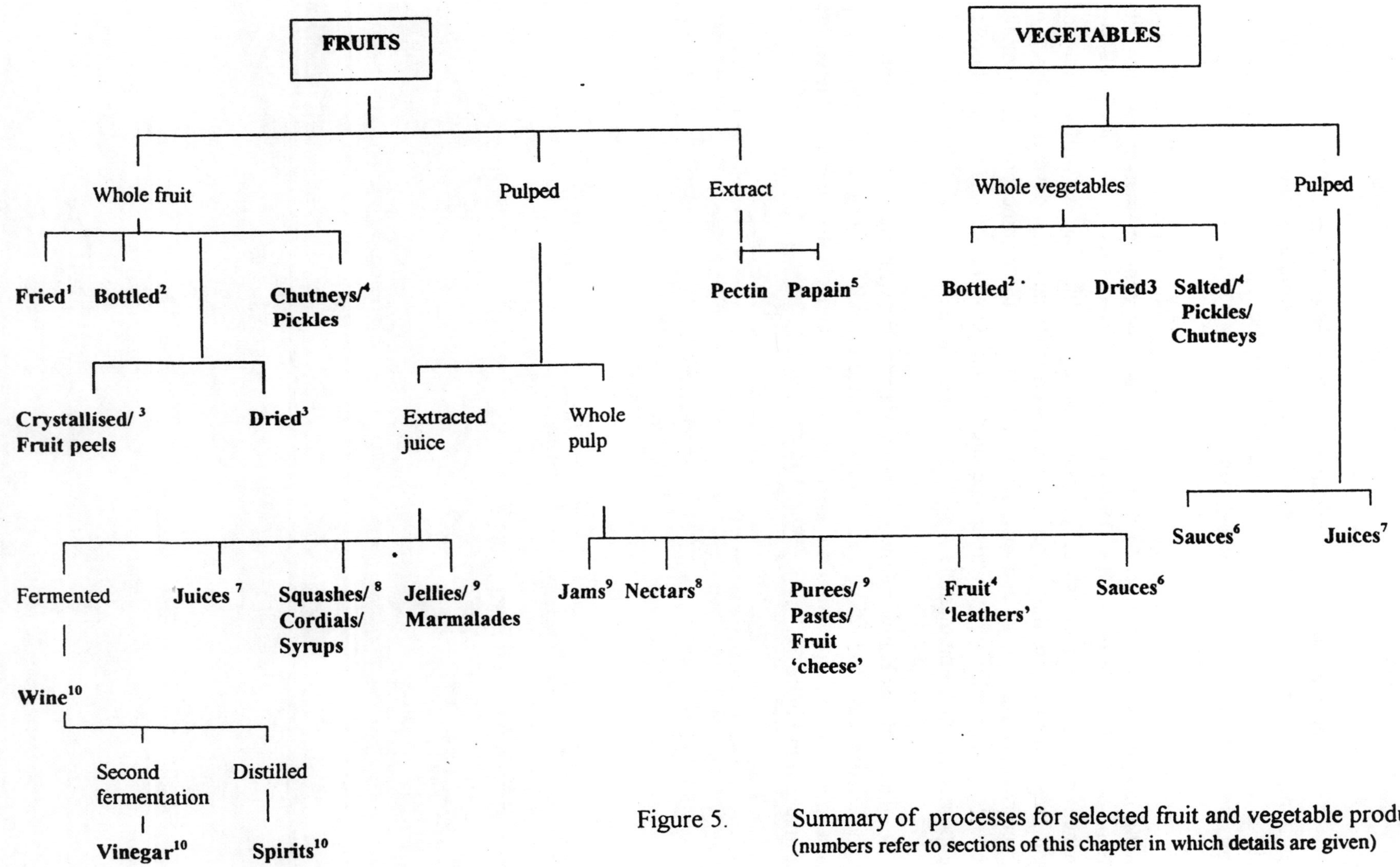

Figure 5. Summary of processes for selected fruit and vegetable products
(numbers refer to sections of this chapter in which details are given)

2.2 Selecting Products and Production Methods

2.2.1 Fried Products

A small number of starchy fruits, including jakfruit, breadfruit and banana are fried and eaten as snackfoods. Heat during frying destroys enzymes and micro-organisms and if sufficient moisture is removed and the product is packaged, it can have a shelf life of several weeks. The most common example of this type of snack-food is banana chips and the process chart (Figure 6) shows their production.

The stages in processing are described in the left-hand column and are split into those steps that are essential for all products (on the right) and those that are only needed for some products (on the left).

Additional processing notes

Care is needed to control the temperature of the oil used for frying, not only for safety reasons, as very hot oil can splash onto operators when wet fruit is immersed, but also because of financial and quality considerations. When oil is heated too much, it exceeds its *smoke point* and a blue haze appears above the oil. This is a sign that the oil is breaking down chemically and it will then begin to get more viscous and develop an unpleasant flavour. The flavour is transferred to the product, making it unacceptable. When the oil thickens, more is retained on the product and there is a higher cost in buying more oil than is needed. Too much oil on the product also reduces its shelf life.

2.2.2 Bottled and Canned Products

Bottling and canning are essentially similar processes in that food is filled into a container and heated to destroy enzymes and micro-organisms. Fruits can be packed into jars with a hot, sugar syrup and vegetables can be packed into a hot brine. The filled jars are sealed and pasteurized so that an internal vacuum forms when they are cool. The sealed container then preserves the food by preventing re-contamination and excluding air and sometimes light. Preservation depends on an adequate heat treatment and an air-tight (or 'hermetic') seal. The process for bottled fruit is shown in Figure 8.

Additional processing notes

There are three grades of syrup: a light syrup contains 200g sugar per litre, a medium syrup 400-600 g/l and a heavy syrup 800g/l. The concentration of salt in brine is usually 15g/l. Acidic fruits require relatively mild heating conditions for pasteurization (e.g. 90-100°C for 10-20 minutes) to destroy yeasts and moulds, whereas less acidic vegetables require more severe heat sterilization to destroy food poisoning bacteria (e.g. 121°C for 15-40 minutes, depending on the size of the container). Fruits can also be part-processed and stored until required in sugar syrup or sodium metabisulphite solution (equivalent to 1000 ppm.) to allow production to take place for a larger part of the year.

It is not advisable for inexperienced small scale processors to bottle vegetables unless they are acidified, because of the risk of poisoning from inadequately processed foods. Vegetables can however be processed by pasteurization if the acidity is first adjusted using citric acid or vinegar.

Stage In Process Essential	Optional	Quality Assurance	Equipment Required	Notes
Fruit ↓				Select fruit that is slightly under-ripe and having a firm texture
Peel ↓				
Slice ↓		Thickness of slices	Slicer	Manual slicing is possible but a thinner, more uniform product is achieved using small manual or electric slicing machines.
Wash ↓		To separate slices	Wash tank	Citric acid or sodium metabisulphite can be added to the wash water to better retain the colour of the fruit and prevent darkening.
Drain	Dry			
Fry ↓		To correct colour Check oil quality	Heat source and pan or deep fat fryer	Care is needed when adding fruit to hot oil (at 180-200°C) to prevent splashing. Heat for 5-10 minutes until a uniform colour is produced. The quality of the oil has a direct effect on the shelf life and quality of the product as well as the economics of the process (see text).
Drain/cool ↓			Racks or mesh	Cool to prevent condensation inside the package
Pack ↓		Check that the seal is adequate.	Heat sealer	Ensure that the seal is complete because the product shelf life is substantially reduced if air or moisture enter through a poorly formed seal.
Store				Store away from heat and sunlight to avoid rancidity

Figure 6. - Process chart for production of fried fruits

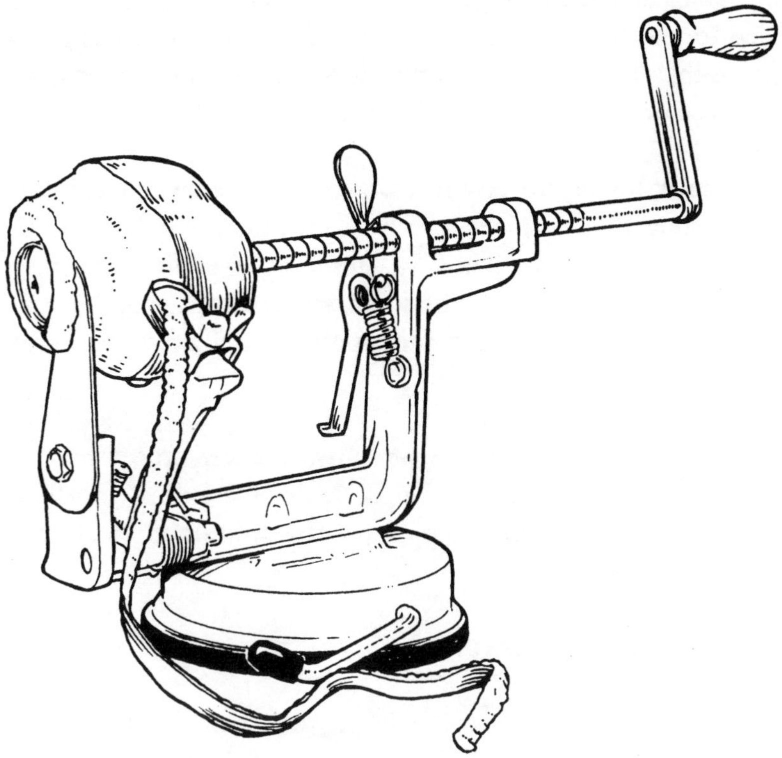

Figure 7. - Small scale peeler for fruits

Canning is not suitable for small scale processing for the following reasons: the time and temperature of canning are critically important and must be carefully controlled. If the cans are under-processed, there is a risk of serious food poisoning and even death from a type of micro-organism named *Clostridium botulinum.* If cans are over-processed, the vegetables lose much of their texture, colour, vitamins and flavour and are not saleable. The establishment of correct heating conditions depends on the type of food, the size and shape of the can and the initial level of contamination of the vegetables. This requires the skills of a qualified food technologist or microbiologist.

When foods are heated in sealed cans during the canning process, the temperature of sterilization is above 100°C and the pressure outside the can must equal that inside, to prevent the cans from exploding. This is achieved using high pressure steam and a strong vessel named a 'retort'. Both steam boiler and retort are expensive and likely to be beyond the means of a small scale processor. Additionally, compressed air is needed to maintain the pressure while cans are being cooled, which together with the necessary controllers, adds to the capital cost of equipment.

Even if cans are available in a particular country, they are usually more expensive than other forms of packaging. Different types of product also require a particular internal lacquer to prevent the metal from corroding when it is in contact with the fruits or vegetables and such lacquers may not always be available. In addition a 'seamer' is needed to seal the lid onto the can and regular checks and maintenance are necessary to ensure that the seam is properly formed. Failures in seams are one of the main causes of spoiled or dangerous canned foods.

It is therefore necessary to ensure that seamer operators are fully trained and experienced in adjusting the machines and a 'seam micrometer' is another necessary capital expense to be able to do this. In summary therefore, canning requires a considerable capital investment, trained and experienced staff, regular maintenance of relatively sophisticated equipment, a regular supply of the correct types of cans and a comparatively high operating expenditure.

Because of the more acidic nature of fruits, a lower processing temperature is adequate and this process is suitable for small scale operations. In all cases, a food technologist should be consulted to advise on process times and conditions for bottled products.

2.2.3 Dried Fruits and Vegetables

Drying removes most of the water from fruits and vegetables to extend their shelf life and to increase their convenience and value. The reduction in weight and bulk also makes transport cheaper and easier although many dried foods are fragile and require packing in boxes to prevent them from being crushed. Different categories of dried foods can be described as *high-volume, lower-value* crops such as staple cereals and *low-volume, higher-value* foods such as dried fruits, vegetables herbs and spices. This second category offers better opportunities for profitable production by small scale processors.

Air dried products are the most common type of dried fruit and vegetables and other more expensive methods, such as freeze drying, are not considered in this book. Some products may be blanched or sulphured/sulphited to protect their natural colour and help preserve them. Crystallized fruits, peels for marmalade and cake production and osmotically dried fruits (known as 'osmasol' products when dried in a solar dryer) are fruit pieces that are soaked in hot concentrated sugar syrups to extract some of the water before air drying.

STAGE IN PROCESS Essential Optional	QUALITY ASSURANCE	EQUIPMENT REQUIRED	NOTES
Harvest fruit ↓			Mature but not over-ripe fruits and vegetables, harvested as carefully as possible to reduce bruising etc.
Wash ↓	To remove contaminants	Wash tanks	Use clean water to remove stones, leaves pesticide residues or soil
Sort/ grade	Essential to produce uniform quality products	Sorting table	Sort by hand to produce uniform sized pieces if whole fruits & vegetables are used and uniform colour or maturity.
Peel	Check that all traces of peel are removed	Knives, peelers	Usually done by hand using sharp stainless steel knives, but small peeling machines are available for some types of fruits (Figure 7).
Cut/slice/core			To reduce the size of pieces to fit into jars and allow faster heat treatment.
Acid dip	Check for uniform sized pieces	Knives, fruit choppers, cutters, slicing or dicing machines	Dipping in 2% citric acid, lemon or lime juice for 5-10 minutes. Is sometimes used to prevent browning of light coloured fruits
↓ Mix ← Sugar syrup ↓	Check acid concentration Check syrup concentration and temperature		Hot fill into pre-sterilised hot jars with 30-50% sugar syrup, heated to 90°C and filtered through a fine muslin bag. Jars should be sterilised by heating to boiling in water for 10 minutes. Re-used jars should be carefully washed.
Fill & Seal ← Jars & lids ↓	Check for correct fill-weight and properly sealed lid	Weighing scales or scoops, heater, filter bag, thermometer, (refractometer optional)	
Pasteurise ↓	Check temperature and time of heating	Boiling water-bath	Heat gently in boiling water for 10-20 minutes, depending on size of the jar
Cool ↓		Cooler (optional)	Cool to room temperature in air or using a water cooler
Label ← Labels ↓ Store	Check label quality & storage conditions	By hand or with label applicator	Check that correct type of label is used and that they are correctly aligned. Store in cool dry place away from sunlight

Figure 8. - Process chart for bottled fruits

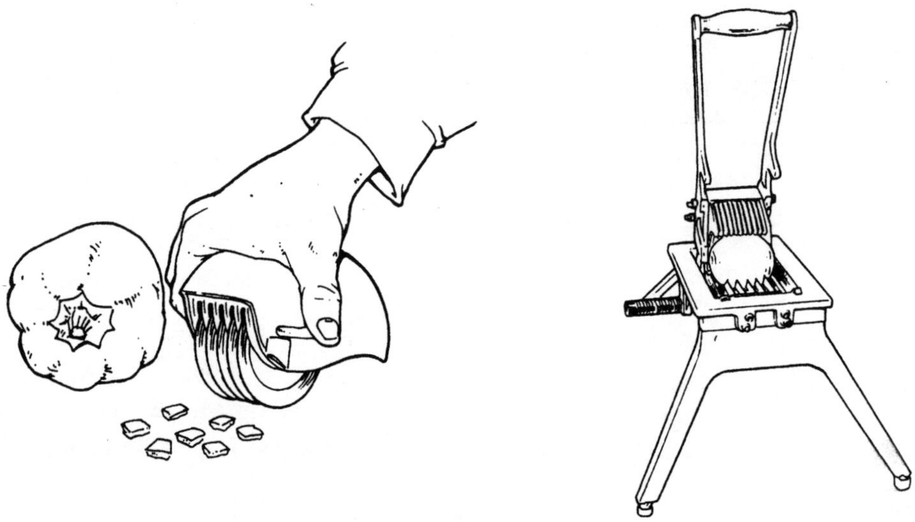

Figure 9. - Examples of fruit cutters suitable for small scale production

The preparation procedures needed for dried fruits and vegetables are summarized in Table 6 and production methods are shown in the Process Chart (Figure 10).

Some vegetables and a few fruits such as limes may also be salted before drying. In this case the high salt concentration preserves the food by both drawing out water by osmosis and by the anti-microbial properties of the salt. Salt tolerant micro-organisms begin to grow while the product is sun dried and these produce acids and characteristic flavours, High salt concentrations also prevent the action of some enzymes, which would cause a loss in quality of the dried food during storage. Vegetables must be washed to lower the salt concentration before they are eaten (vegetables that are salted but not dried are described in Section 2.2.4).

Additional processing notes

Fruits and vegetables must be carefully selected before drying. If fruits in particular are over-ripe they are easily damaged and may be difficult to dry. If they are under-ripe, they have a poorer flavour, colour and appearance. Care and attention to hygiene are essential because any bacteria or moulds that contaminate vegetables before drying are likely to survive on the dried food. The temperature of drying is not high enough to kill them and when the food is re-hydrated, they can grow again and cause food poisoning.

Stage in Process Essential / Optional	Quality Assurance	Equipment Required	Notes
Harvest	Check for full maturity but not over-ripe		Harvest as carefully as possible to reduce bruising, etc.
Wash	To remove contaminants	Wash tanks or special washers.	Use clean water to remove stones, leaves, pesticide residues or soil
Sort/grade			Sort by hand to get similar colour or maturity. Uniform maturity is important to get a similar drying time for all pieces.
Peel (optional)	Essential to produce uniform quality products. Check that all traces of peel are removed.	Knives, peelers	Peeling increases the rate of drying because peel prevents moisture leaving the food. It is usually done by hand using sharp stainless steel knives but small peeling machines are available for some types of fruits (Figure 7).
Cut/slice/core (optional)	Check for uniform sized pieces	Fruit & vegetable choppers, cutters, slicing or dicing machines.	To reduce the size of pieces and allow faster blanching, sulphur dioxide treatment and drying.
Pulp (optional)	Uniform pulp, no fruit pieces	Pulpers, liquidisers, steamers or, at a larger scale, pulper-finishers	Only for fruit leathers. Small scale pulpers are preferable, but this stage can also be done by hand or using steamers. The pulp is spread on polythene sheets for drying.
Blanch (optional)	Check blancher water temperature, time of heating and concentration of any salts added	Boiling pan, heat source, wire basket or steamer.	More often used for vegetables. Care is needed to prevent blanched foods from being recontaminated before drying.
Acid dip (optional)	Check concentration of acid.	Weighing scales or scoops, food grade plastic tank.	Dipping in 2% citric acid, lemon or lime juice for 5-10 minutes is sometimes used to prevent browning of light coloured fruits and vegetables.
Treatment with sulphur dioxide (optional)	Check weight of sulphur or concentration of sulphite and time of treatment	Weighing scales or scoops, sulphuring cabinet or food-grade plastic tank.	Optional for some fruits and vegetables to reduće browing during drying. Can be done by either 1) sulphuring (exposing fruit to burning sulphur in a sulphuring cabinet ir 2) treatment in a sulphite dip. Care is needed as sulphur dioxide causes coughing and eye irritation.
Syruping (optional)	Check syrup concentration and temperature.	Weighing scales or scoops, boiling pan, heater, food-grade tank, muslin cloth filter.	Only used for crystallized, osmasol dried fruits or to make peels for marmalade (Section 2.2.10). In the simplest method, fruit is boiled in 60% sugar syrup for 10-15 minutes and then soaked in the same syrup for 12-18 hours (also Figure 11)
Dry	Check rate and temperature of drying, check for mould growth and insect contamination.	Sun drying, solar dryer, fuel-fired dryer or electric dryer.	The rate of drying depends on the temperature, humidity and speed of the air, the type of dryer and the size of the food pieces. Generally fruits and vegetables are dried at 40-60°C. Different types of dryer are described in the text.
Pack ← Bags	Check fill-weight and correctly sealed pack	Electric heat sealer for plastic bags (Figure 46	Required to keep product pieces together. If product risks gaining moisture, package must be moisture-proof - see text.
Label and store	Check that label and storage conditions are correct.		Correct label for type of product and store in a cool dry place away from sunlight. Fragile foods may need protection from crushing.

Figure 10. - Process chart for dried fruits and vegetables

Blanching

Blanching destroys enzymes and prevents changes in colour, flavour and texture during storage. However, by itself, it does not preserve the food and vegetables must therefore be further processed by drying to achieve a long shelf life. Vegetables are blanched by heating in hot water or steam for a short time (Table 7) and then cooled on trays. For production at a small scale, vegetables can be placed in a wire basket and immersed in boiling water (Figure 42).

In steam blanching, vegetables are placed in a strainer and this is then fitted over a pan of boiling water and covered with a lid to prevent the steam escaping. Steaming takes a few minutes longer than water blanching, but has the advantage of retaining more nutrients as they are not lost into the water.

There are optional chemical treatments that help to retain the colour and texture of some dried fruits and vegetables. For example, the bright green colour of leafy vegetables, peas etc. can be retained by adding sodium bicarbonate to blancher water and the texture of some vegetables, such as okra and green beans, can be maintained by blanching in a calcium chloride solution. Both chemicals are usually available from pharmacies in major towns.

Sulphuring and sulphiting

For most fruits, 350-400g sulphur are used per 100kg fruit, burning for 1-3 hours. Sulphur dioxide prevents browning in foods such as apple, apricot and coconut, although it should not be used with red fruits as it bleaches the colour. Sulphuring (using sulphur dioxide gas) is achieved by exposing pieces of cut or shredded fruits to burning sulphur in a sulphuring cabinet. The amount of sulphur used and the time of exposure depend on the type of fruit, its moisture content and limits placed by law in some countries on the residual amounts of sulphur dioxide in the final product (Section 2.4.2) or by commercial limits set by importers. This should be checked with a local Bureau of Standards. There is an increasing consumer resistance to sulphited fruits in some industrialised countries and if the product is considered for export the local Export Development Board or import agents should be consulted.

In sulphiting, the sulphur dioxide is dissolved in water, rather than as the gas used in the sulphuring process. Sodium sulphite, sodium metabisulphite or potassium metabisulphite are made into solutions, either by adding one of them to the blanching water or more often, by soaking the food for 5-10 minutes in a sulphite dip. About two thirds of the weight of sodium metabisulphite is formed as sulphur dioxide when it is dissolved in water. For example, to form a 0.001% solution which is equivalent to 1000 ppm, 1.5g is dissolved in a litre of water to form 1g of sulphur dioxide per litre. At this concentration, sulphiting can also be used as a method of intermediate storage of fruits to spread production over several months throughout the year.

	Cleaning	Sorting/ Grading	Peeling	Size reduction	Blanching	Salting	Sulphur dioxide treatment
Fruits	Needed for all fruits	Important for all fruits	Required for most products	Cutting larger fruits for faster drying	Not usually done for fruits	Very few	For some fruits to prevent browning
Vegetables	Needed for all vegetables	Important for all vegetables	Used for some vegetables	Slicing for larger vegetables or shredding cabbage	Commonl y used for softening and preventing browning	For some salted vegetable products (e.g. cabbage)	Used with a few light coloured vegetables to prevent browning

Table 6. - Preparation procedures for fruits and vegetables before drying

Syrup pre-treatment

This method can be used to remove up to half of the water in fruit and is therefore a cheap way of increasing the production rate of a dryer or for part-processing fruits for intermediate storage so that production can be extended throughout the year. In general the method gives good retention of colour in the dried food and produces a sweeter, blander tasting product. However, acids are also removed from fruits during the process and the lower acidity of the product may allow mould growth if the food is not properly dried and packaged.

Food	Blanching time (minutes) using:	
	Steam	Water
Leafy vegetables,	2 - 2.5	1.5
Sliced beans	2 - 2.5	1.5 - 2
Squashes	2.5	1.5 - 2
Cabbage	2.5	5 - 2
Peas	3	2
Carrots	3 - 3.5	3.5
Cauliflower	4 - 5	3 - 4
Potatoes	6 - 8	5 - 6

Table 7. - Blanching times for different vegetables

In a more complex method to that described in Figure 10, fruit is first boiled in 20% syrup and then soaked overnight. The fruit is then strained from the syrup and transferred each day to 40% and 50% syrups in turn (Figure 11), with optional boiling for 10 minutes at each transfer. After soaking, the syrup is diluted to approximately half of the original concentration. Each day the most dilute syrup (10%) is used for other products and a new 60% syrup is made up. The advantages of this method include re-use of sugar syrups and a softer texture in the final product. In Figure 11, the containers therefore circulate 'backwards' as they contain more dilute syrup. Some producers have even more stages in the process and may transfer fruit into increasing sugar concentrations each day for up to fourteen days. This results in a succulent, soft texture in the final product.

Types of dryers

The higher value of dried fruit and vegetable products, compared for example to cereals crops, may justify the higher capital investment in a fuel-fired dryer or electric dryer and the extra operating costs for the fuel or electricity. These types of dryer allow higher drying rates and greater control over drying conditions than do solar or sun drying and they can therefore result in a higher product quality. However it is necessary to make a careful assessment of the expected increase in income from better quality products compared to the additional expense, to make sure that this type of dryer is cost-effective (see also Section 2.3).

Sun drying is only possible in areas where, in an average year, the weather allows foods to be fully dried immediately after harvest (see Figure 3, seasonality chart. The main advantages of sun drying are the low capital and operating costs and the fact that little expertise is required.

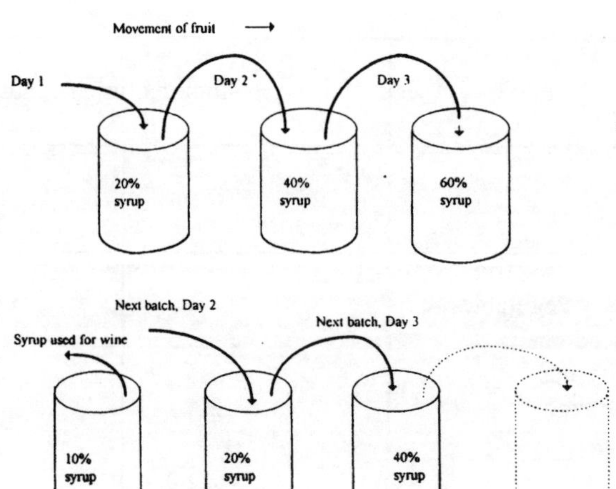

Figure 11. - Process for making crystallised (or candied) fruits

The main problems with this method are as follows:

- contamination, theft or damage by birds, rats or insects
- slow or intermittent drying and no protection from rain or dew that wets the product, encourages mould growth and may result in a relatively high final moisture content
- low and variable quality of products due to over- or under-drying
- large areas of land needed for the shallow layers of food
- laborious because the crop must be turned, moved if it rains and animals must be kept away
- direct exposure to sunlight reduces the quality (colour and vitamin content) of some fruits and green vegetables.

The quality of sun dried foods can be improved by reducing the size of pieces to get faster drying and by drying on raised platforms, covered with cloth or netting to protect against animals and insects (also Section 1.2.1).

Solar drying has been studied in detail by scientists in many countries for many years (see Bibliography), but it is not yet widely used commercially. The main applications to date are in Bangladesh for desiccated coconut, in Guatemala for herbal teas and in Uganda for dried fruits for export. Details of equipment for solar drying and fuel-fired drying are given in Section 2.5.4.

Packaging

If the climate is dry, it may not be necessary to package dried foods as they will not pick up moisture from the air. However a humid climate is likely to result in dried foods gaining moisture and going mouldy. The stability of dried foods depends not only on the humidity of the air at which a food neither gains nor loses weight (the 'Equilibrium Relative Humidity'), but also on the type of food. Different foods can be grouped according to their ability to absorb moisture from the air. The two groups are *hygroscopic*, which absorb moisture easily and *non-hygroscopic*, which do not absorb moisture. The classic example is salt and pepper, where salt is very hygroscopic and pepper is non-hygroscopic, but similar examples exist for fruit and vegetable products. This difference determines the packaging requirement for different fruit and vegetable products. The moisture content at which a food is stable is known as the *Equilibrium Moisture Content* and examples of this for different fruits and vegetables are shown in Table 8, together with the packaging requirement for different groups of foods.

Dried fruits and vegetables are usually packaged in one of the many different types of plastic film. The selection of the correct type of packaging material depends on a complex mix of considerations which include:

- the temperature and humidity of the air in which the product is stored
- the capacity of the product to pick up moisture from the air
- reactions within the product caused by air or sunlight during storage
- the expected shelf life
- marketing considerations (Section 2.8.3)
- cost and availability of different packaging materials.

In general, although thin polythene film is usually the cheapest and most widely available material, it is only suitable for storing dried fruits and vegetables for a short time before they pick up moisture, soften and go mouldy. Polypropylene has better barrier properties and therefore gives a longer shelf life, but it is usually more expensive and it may not be available in many countries. Other more complex films, such as laminated films made from polythene and aluminium foil, offer much better protection to dried foods, but are considerably more expensive and more difficult to find in developing countries. Details of this complex area are given in publications in the bibliography and discussed further in Section 2.5.5.

Food	Moisture content (%)	Degree of protection required
Fresh fruit and vegetables	75-85	Package to prevent moisture loss
Marmalade	35	Non-hygroscopic:- Minimum protection or no packaging required
Raisins	7	
Fruit sweets	3	Hygroscopic:- Package to prevent moisture uptake
Potato crisps	1.5	

Table 8. - Moisture contents at which selected foods are stable and packaging requirements for each group
(Adapted from *Food Processing Technology* by Fellows)

Most dried foods also need a sturdy box or carton to both prevent crushing and to exclude light which causes loss of colour and development of off-flavours during storage. The properties of different packaging materials for dried foods are shown in Table 9.

From the table, it can be seden that some types of packages provide good protection against air and moisture pickup for example, whereas other protect against light, crushing, etc. It is therefore common for dried foods to be packed in airtight and moisture-proof bags, which are then placed in an outer container to protect against light, crushing, etc.

Type of Packaging	Protection provided against:							
	Moisture	Light	Air & odours	Heat	Micro-organisms	Dust	Crushing	Animals & insects
Clear glass	3	1	3	2	3	3	3	3
Coloured glass	3	2	3	2	3	3	3	3
Ceramic pot	3	3	3	3	3	3	3	3
Metal tin	3	3	3	1	3	3	3	3
Metal foil	2	3	2	1	2	3	1	1
Plastic pot	3	3	3	2	3	3	2	2
Wooden chest	2	3	1	3	1	3	3	2
Paper board (cardboard) box	1	3	1	3	1	3	2	1
Fibreboard drum	1	3	1	3	2	3	3	2
Paper bag	1	2	1	1	1	2	1	1
Polythene film	2	1	1	1	2	3	1	1
Cellulose film	3	1	3	1	2	3	1	1
Polypropylene film	3	1	3	1	2	3	1	1
Cotton or jute sack	1	2	1	1	1	2	1	1

Notes: It is assumed that all packs are properly sealed. 1 = poor protection, 2= fair protection, 3 = good production.

Table 9 - Properties of packaging materials for dried fruits and vegetables (Adapted from *Food Processing Technology,* by Fellows and *Appropriate Food Packaging,* by Fellows and Axtell)

2.2.4 Chutneys, Pickles and Salted Vegetables

Chutneys

Chutneys are thick, jam-like mixtures made from a variety of fruits and vegetables, sugar, spices and sometimes vinegar. Any edible sour fruit can be used as a base for a chutney, to complement the sweet taste from the sugar. The high sugar content has a preservative effect and vinegar addition is not always necessary, depending on the natural acidity and maturity of the fruits that are used. Most products are boiled, which not only produces a caramelised syrup and alters the taste, colour and thickness, but also pasteurises the product and thus adds to the preservative action of the sugar and acids. Other products are allowed to ferment naturally and the acids produced by a mixture of bacteria preserve the product. Depending on the types of spices that are added, these may also have a preservative effect, in addition to their contribution to flavour. Fruits can also be sulphited as a method of intermediate storage to spread production over several months throughout the year. The process chart (Figure 13) uses a formulation for mango chutney as a typical product of this type.

Figure 12. - Using a hand-held refractometer to check sugar content
(Courtesy of Midway Technology Ltd.)

Additional processing notes

Natural acids from the fruit, from vinegar or those produced by fermentation, together with the high sugar content, are used to preserve the chutney after a jar has been opened. A correct balance between the levels of sugar and acid is required to prevent mould growth and a *Preservation Index* can be used to calculate the amounts of ingredients to be added. Alternatively, when sugar is the main ingredient or the product is boiled, a refractometer (Figure 12) can be used to check that the final sugar content of the syrup is 68-70%. Sugar is added before heating if a dark product is required or towards the end of boiling to produce a light coloured product.

The Preservation Index is a measure of the preserving power of combinations of acid and sugar (sugar is measured as 'total solids'). This is used to assess whether a chutney or pickle is safe from food spoilage and food poisoning micro-organisms. The value can be calculated as follows:

$$\frac{\textbf{Total acidity x 100}}{\textbf{(100 – total solids)}} = \text{not less than } 3.6\%$$

Details of how to measure acidity and total solids are given in Section 2.7.2. However, if a manufacturer has no access to basic laboratory equipment or is not sure how to carry out the calculation, it is best to take a sample of product to a Bureau of Standards, University Food Science Department or food testing laboratory which can analyse it and recommend adjustments to the recipe if necessary.

Pickles

Vegetables such as cucumber, cabbage, olive and onion are fermented by lactic acid bacteria which can grow in low concentrations of salt. The bacteria ferment sugars in the food to form lactic acid, which then prevents the growth of food poisoning bacteria and moulds or other spoilage micro-organisms. The amount of added salt controls the type and rate of the fermentation. If for example 2-5% salt is used, a natural sequence of different types of bacteria produce the lactic acid. If higher concentrations of salt (up to 16%) are used, a different product called 'salt stock' pickle is produced, which is preserved by the salt and not by fermentation. Fruits and vegetables can be preserved in this way as a method of intermediate storage to spread production over several months throughout the year.

Sometimes, sugar is added to increase the rate of fermentation or to make the product sweeter. Alternatively, vegetables may be packed in vinegar (acetic acid), salt and sometimes sugar to produce a variety of pickled products. Because these vegetables are not fermented they have a different flavour and texture. They are usually pasteurised. Sweet pickles are made from single fruits or mixtures of fruits and vegetables. They are preserved by the combined action of lactic or acetic acid, sugar and in some cases added spices. A summary of the different types is shown in Table 10 and a process chart (Figure 14) describes their production.

Stage In Process Essential / Optional	Quality Assurance	Equipment Required	Notes
Harvest fruit	Check for full maturity but not over-ripe		Picked carefully to reduce bruising etc. which would cause spoilage and loss of quality.
↓ Wash		Wash tank	Use clean water to remove stones, leaves or soil
↓ Sort/ grade	Essential to produce uniform quality products		Sort by hand to get similar colour or maturity.
Peel (optional)	Check that all traces of peel are removed	Knives, peelers	Most fruits require peeling, but some chutneys include un-peeled fruits. By hand using sharp stainless steel knives or small peeling machines for some types of fruits (Figure. 7)
Cut/slice/core (optional)	Check for uniform sized pieces	Knives, slicers, dicers	Most fruits are cut into small strips. It is important to reduce the size of pieces to obtain uniform mixtures of similar sized fruits and/or vegetables. Cutting also allows faster penetration of sugar. This may be done by hand, but small scale equipment is available to increase production rates (Figure 9).
↓ Mix ← Spices, sugar vinegar		Mixing bowl, scales	For each 500g of mango, spices such as ginger (0.5g), mustard (0.5g), chilli (1g), garlic (0.5g) and salt (1g) are washed, ground and mixed with vinegar (250ml) and sugar (300-500g). In some recipes, 7% oil is added.
Heat (optional)	Check time and temperature	Heat source, boiling pan, (refractometer optional)	Boil until sufficiently thick and concentrated. Heat destroys enzymes and micro-organisms which together with high sugar level, spices & acid from the vinegar, helps to preserve product.
Ferment (optional)	Check temperature and absence of insects	Covered pans	Natural fermentation over 2-3 days produces acids to assist preservation. Traditionally such products were not pasteurised.
↓ Fill & Seal ← Jars/ lids	Check fill-weight and correctly sealed pack	Funnel or paste filler, scales	Packaging contains the contents and prevents re-contamination. Hot fill into pre-sterilised jars and hand seal lids - see text for alternatives.
↓ Cool, Label & Store	Check label & storage conditions		In boxes, on racks, with a cool dry atmosphere.

Figure 13. - Process chart for chutney (e.g. mango chutney)

Salted vegetables

Salted vegetables are made by building up alternate layers of chopped or shredded vegetable such as cabbage, with layers of salt in a sealed drum (see also Section 1.2.1). The salt has two preservative actions: it draws out water from the vegetables by osmosis to form a concentrated brine in the base of the drum; and salt also has a direct anti-microbial action. The high levels of salt are reduced by washing the products before they are eaten.

Product	Salt	Sugar	Vinegar	Process
Fermented sweet pickle	5% then 3%	1-2% then 3%	0 then 5%	Fermented for 1-2 weeks then repack in vinegar + salt+ sugar (optional pasteurization)
Fermented sour pickle	5% then 3%	0 then 0	0 then 5%	Ferment for 1-2 weeks then repack in vinegar + salt (optional pasteurization)
Unfermented pickle	3%	1%	5%	Pack straight away and pasteurize
Salt-stock pickle	15%	0	0	Store until required. Then wash out salt and repack as unfermented pickle

Table 10. - Examples of different types of fruit or vegetable pickles

Additional processing notes

Pickles which have an adequate Preservation Index do not need to be pasteurised. However as an additional measure to prevent spoilage, they can be pasteurised or the sugar/salt/vinegar mixture can be heated, added to the vegetables and the jars filled while product is still hot. In this way the hot product forms a partial vacuum in the jar when it cools and further aids preservation.

Glass jars are the most commonly used packaging material, but if a shorter shelf life is expected, pickles may also be packed in small quantities in polythene pouches and sealed with an electric heat sealer. To avoid seepage of product, which can damage paper labels and make the package unattractive, a double pouch can be used comprising an inner pack that contains the product and an outer pouch with a label between the two.

Stage In Process Essential Optional	Quality Assurance	Equipment Required	Notes
Harvest vegetables ↓	Check for full maturity but not over-ripe		Cucumber, cabbage, courgette or other types of vegetable, picked carefully to reduce splitting or bruising.
Wash ↓		Wash•tank	Use clean water to remove stones, leaves or soil.
Sort/ grade	Essential to produce uniform quality products		Sort by hand to get similar colour, size, shape or maturity.
Trim (optional)		Knives	Remove discoloured leaves, stems or other unwanted parts
Cut/slice/core (optional) ↓	Check for uniform sized pieces	Knives, slicers, dicers	For some types (e.g. shredded cabbage, sliced limes) but small cucumbers and courgettes are usually pickled whole. Cutting allows faster penetration of pickling liquor.
Mix ← Salt, sugar, vinegar	Check weights of ingredients	Food grade plastic tank, scales, (pH meter, brine hydrometer optional)	Different mixtures are used depending upon tradition, local tastes and the different product requirements (see Table 10)
Ferment (optional)	Check for contamination by insects	In fermentation tank	Natural fermentation for 3-7 days/up to 18 months produces acids to assist preservation. Traditionally such products were not pasteurised.
Heat (optional) ↓	Check time and temperature	Heat source, boiling pan, (refractometer optional)	Pasteurise for 10-15mins at 90-100°C destroys enzymes and micro-organisms. Together with acid from the vinegar or fermentation, this helps to preserve product.
Fill & Seal ← Jars/ lids ↓	Check fill-weight and correctly sealed pack	Filler, scales	Remove from barrels/drums and re-pack into pre-sterilised jars. Hot-fill with new brine/vinegar to cover vegetables. Final salt and acetic acid content as shown in Table 10.
Cool, Label & Store	Check that label & storage conditions are correct		Store in boxes on racks, with a cool dry atmosphere.

Figure 14. - Process chart for pickled vegetables

2.2.5 Pectin and Papain

Pectin

Pectin is a component of nearly all fruits and vegetables and can be extracted and used in food processing to form the characteristic gel in jams and marmalade. The richest sources of pectin are the peels of citrus fruits such as lime, lemon, orange and passion fruit or the pulp of apple (known as 'pomace' after juice has been extracted). Commercially, pectin is available as either a light brown powder or as a dark liquid concentrate. It is stable if stored in cool, dry place and it will lose only about 2% of its gelling power per year.

Stage in Process		Quality Assurance	Equipment Required	Notes
Essential	Optional			
Fruit		No special checks, but mouldy fruit should be discarded		Under-ripe fruit has a greater amount of pectin than over-ripe fruit.
Wash			Wash tank.	To remove pesticide residues, leaves, etc.
	Peel		Knives	Remove citrus peels and use for extraction. Other fruits are not peeled.
Cut		Check that pieces are small and uniform.	Knives, slicers	Peels are cut into thin slices (e.g. 1cm wide) and other fruits are coarsely chopped.
Mix	← Water			Add water - approximately 8 times the weight of fruit.
Heat		Check the temperature and time of heating.	Boiling pan, heat source.	Heat gently to simmering and hold at that temperature for 20-30 minutes.
	Concentrate			Longer heating removes more water and concentrates the pectin so that it can be stored for a few days. However, over-long heating degrades the pectin and it loses some of its gelling power.
use in jam making process				

Figure 15. - Process chart for pectin extraction

There are two main types of pectin:

1) high methoxul (HM) pectins that form gels in high solds jams (above 55% solids) in a pH range of 2.0-3.5; and

2) low methoxyl (LM) pectins, which do not need sugar or acid to form a gel, but instead use calcium salts.

These form a gel with a wide range of solids (10-80%) within broader pH range of 2.5-6.5. They are used mainly for spreads or for gelling agents in milk products. However, there are a large number of different types of pectin within each group, such as 'rapid set' and 'slow set' pectins, that are made for different applications and it is necessary to specify carefully the type required when ordering from a supplier.

It is difficult to make specific grades of pectin or the powdered products at a small scale. Where attempts have been made to set up small commercial pectin units, the process has not been financially viable due to the expense of recovering solvents that are used to precipitate the pectin before drying. However, it is possible for jam-makers to increase the amount of crude (unrefined) pectin in their products by extracting it with water without using solvents (Figure 15). This is useful if the producer is making jam from fruits such as melon that are naturally low in pectin. When using pectin to make set preserves, such as jams, jellies etc., it is important that the pH is within the range of 3.0-3.5 and that they are boiled to at least 68% solids content. Powdered pectin is added to fruit pulp at 3-6g per kg of final product, but it should first be mixed with about five times its weight of sugar to prevent lumps forming when it is added to the pulp or juice (see also Section 2.2.9).

Papain

Papain is an enzyme that is found in the skin of unripe papaya fruits. It breaks down proteins and finds widespread applications in industrialised countries for meat tenderising and in brewing. Advances in biotechnology during the last twenty years resulted in a synthetic papain that was cheaper than the natural enzyme and as a result there was a decline in the market for natural papain. However, in recent years, there has again been increasing interest in the natural product and it can now achieve a higher value than before.

For profitable production, it is usually necessary to plant papaya trees in orchards, rather than collecting it from widely scattered trees which increases collection costs. Crude (unrefined) papain is normally produced because the refining technology is too expensive for most small-scale entrepreneurs. The process of papain extraction involves making shallow cuts in the skin of unripe fruits while they are still on the tree. The skin then oozes a white, sticky liquid which is collected by scraping the fruit each day for several weeks. This 'latex' is then spread out in shallow trays and sun-dried until it becomes brittle. The crude papain is packaged and exported to refiners. Because papain is an enzyme that attacks proteins, the crude latex should be handled with gloves to avoid damage to the operator's skin.

2.2.6 Sauces

Sauces are thick viscous liquids, made from pulped fruit and/or vegetables with the addition of salt, sugar, spices and vinegar. They are pasteurised to give the required shelf life, but the basic principle of preservation is the use of vinegar, which inhibits the growth of spoilage and food poisoning micro-organisms. Other ingredients such as salt and sugar contribute to the preservative effect and the correct Preservation Index (Section 2.2.4) ensures that the product does not spoil after opening and can be used a little at a time. Some may contain a preservative such as sodium benzoate, but this is not necessary if an adequate Preservation Index is achieved. Sauces can be made from almost any combination of fruit or vegetables, but in practice the market in many countries is dominated by tomato sauce, chilli sauce and to a lesser extent, mixed fruit sauces such as 'Worcester' sauce, which contains apples and dates in addition to tomatoes. Depending on the scale of production, pulping and sieving out seeds and skins can be done by hand or using special pulper-finisher machines (Figure 16). The process for making sauce is outlined in the process chart below (Figure 17), using tomato sauce as an example.

Additional processing notes

Similarly, at a small scale, sauces can be made using simple open boiling pans, provided that care is taken to heat slowly with constant stirring to avoid localised burning of the product, especially at the end of heating. At a larger scale, processing is done using steam heated, stainless steel 'double jacketed' pans (Figure 18).

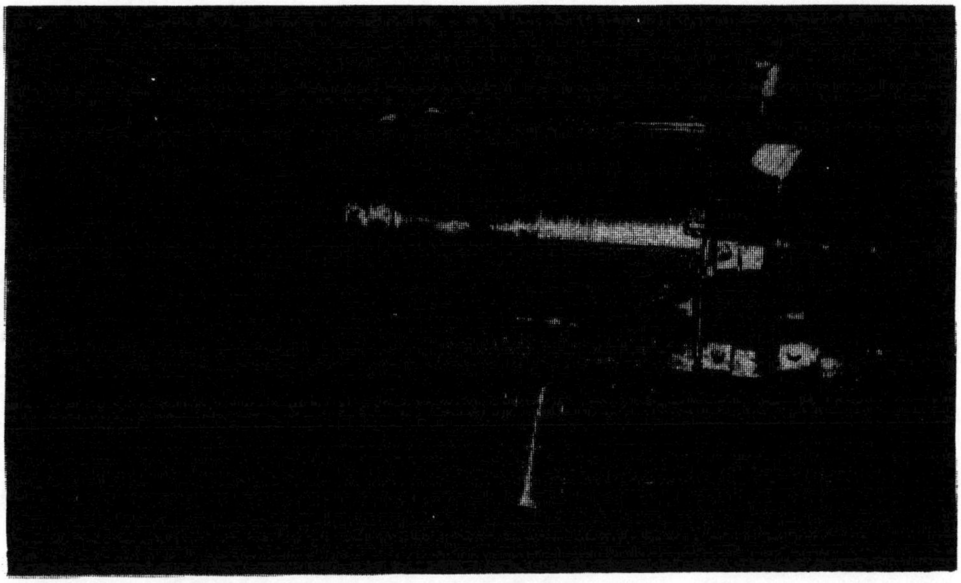

Figure 16. - Pulper-finisher used to extract fruit pulp without seeds or skins

Stage in Process Essential / Optional	Quality Assurance	Equipment Required	Notes
Harvest fruit ↓	Check for full maturity.		Mature but not over-rip fruit (of the 'plum'tomato variety) are picked carefully to reduce puncturing, bruising, etc., which would cause spoilage and loss of quality. Transport in stackable boxes to avoid crushing.
Wash ↓		Wash tank.	Use clean water to remove dust, leaves, pesticide residues, etc.
Sort ↓	Essential to produce uniform quality products.		Sort by hand for similar colour, discard under-ripe, mouldy or discoloured fruits.
Heat/Peel	Check temperature and time of heating.	Boiling pan, wire basket (pulper-finisher optional)	Tomatoes are placed into a wire basket and heated in boiling water for 10 minutes. They are then removed and peeled by hand. Alternative, use a pulper finisher.
Pulp or Chop ↓	Check for uniform pulp.	Mesh screen.	Chop finely by hand if heat peeled. Mesh screen not needed if pulper-finisher used.
(Optional) Filter	Check all seeds removed.		
Mix ← Spices sugar, vinegar salt ↓	Check correct formulation used.	Weighing scales or scoops, muslin bag.	A typical formulation per kg of tomatoes is 10g salt, 200ml vinegar and 80g sugar. Herbs and spices, such as 1g each of sinnamon, ground cloves, Allspice, and cayenne pepper are placed into a tied muslin bag and this is submerged in the pulp.
Heat ↓	Check time and temperature (measurement of solids content optional)	Boiling pan, heat source, scales or scoops, refractometer.	The mixture is heated to 80-90°C until thick, usually within 20 min. Vinegar is added towards the end of the boiling to reduce the amount of acetic acid that is lost by evaporation.
Fill & Seal ← bottles & caps ↓	Check fill-weight and correctly sealed pack.	Funnel or paste filer, scales.	Hot fill into pre-sterilized bottles and hand seal lids - see text for alternatives.
Heat ↓	Check time and temp., of heating	Stainless steel boiling pan.	Pasteurize at 88°C for 20-30 min. Depending on the size of the bottle.
Cool, Label & Store	Check label	(Label appalicator optional)	In boxes on racks, in a cool dry atmosphere, away from sunlight.

Figure 17. - Process chart for tomato sauce

2.2.7 Juices

The beverage market is divided into products that are intended to quench thirst, such as juices, nectars and carbonated drinks and those that are drunk on special occasions, such as wines and spirits, where religion or custom permit. Competition from large scale manufacturers is among the fiercest in beverage manufacture and considerable amounts of money are spent on advertising, packaging and sophisticated distribution systems in this sub-sector. Thus beverage manufacture is one of the most difficult for small scale producers to become established and successful. However there is a growing trend in urban centres of some developing countries towards increased juice consumption and this market may become larger in future years.

In this Section, the process for making juice is described first. In subsequent processes for other drinks, the base material, fruit pulp, is used as the starting point in the process. The description of juice manufacture (Figure 19) should therefore be read in conjunction with other sections.

All drinks in this Section contain pulped fruit or juice from a single fruit or a mixture of fruits. Carbonated drinks, which do not contain fruits are not included in this book. Drinks can be divided into those that are drunk immediately after opening and those that are diluted before use. The first group should not contain any preservative but the second group may need a preservative such as sodium benzoate to have a long shelf life after opening. Unopened bottles of both types should have a shelf life of 3-9 months, depending on the storage conditions.

Fruit juices should be pure juice, with nothing added (see Section 2.4.2), except in products such as some types of melon juice, in which the level of acidity needs to be increased by addition of citric acid to give a pH below approximately 3.5-4.0. Preservation is due to pasteurisation and the natural acidity of the juice. The process for juice production is shown in Figure 19.

Additional processing notes

A wide range of drinks can be made from pulped fruit or juice and production can be spread over a larger part of the year by processing a sequence of fruits or by part-processing pulps and storing them in 1000-2000 ppm. sodium metabisulphite solution.

Acidity

pH paper is useful for checking the acidity of some less acidic juices (e.g. melon juice at pH 4.0-5.0). The optimum range is 3.0-4.0 and it may be necessary either to blend juices or adjust the pH with citric acid, to prevent spoilage of less acidic juices.

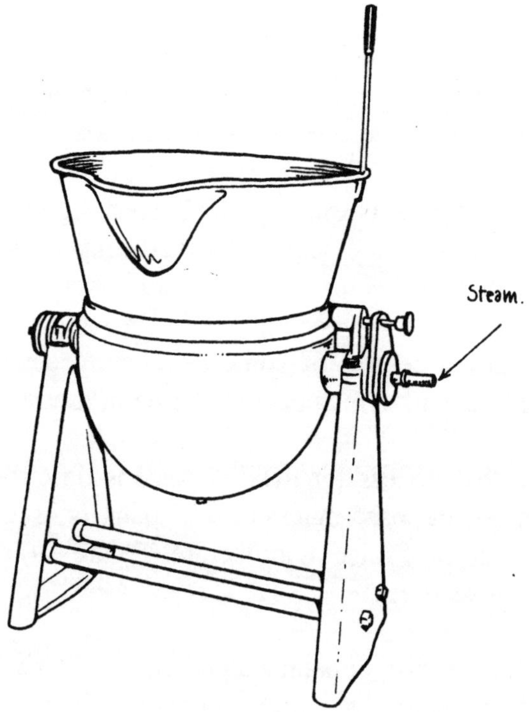

Figure 18. - 'Double jacketed' pan for larger scale boiling

Pulping

Soft fruits such as berries, passion fruit and papaya can be pulped by hand, whereas harder and more fibrous fruits such as pineapple and mango require mechanised pulping. Manually operated pulpers are available for places where there is no electricity and there are also a wide range of powered pulpers and blenders. At higher production rates, pulper-finishers brush fruit through a sieve and separate seeds, skins etc. from the pulp. Juice can also be extracted from fruit using a fruit press or fruit mill or by steaming the fruit. Citrus fruit juices are extracted by reaming the fruit and again comparatively simple equipment is available for this purpose (Figure 45).

Pasteurisation

Drinks need to be pasteurised if they are to have a shelf life of more than a few days. The time and temperature required for pasteurisation depends on the type of

Stage in process: Essential	Optional	Quality Assurance	Equipment Required	Note
Harvest fruit		Check for full maturity.		Fruits should be mature or over-ripe, without mould growth. They should be picked carefully to reduce puncturing, splitting or bruising.
Wash			Wash tank	Use clean water toremove stones, leaves or soil.
Sort/grade		Essential to produce uniform quality products		Sort by hand to get similar colour or maturity and remove all mouldy fruits.
	Cut/slice/core		Knives	For some types of fruit (e.g. pineapple). This stage also depends on whether pulping is done mechanically or by hand. Pulper finishers (Figure 16) are able to separate seeds and skins without cutting, but sieving/ filtering is needed if pulping is done by hand.
Pulp		Check for uniform pulp.		
Filter → Fruit pulp			Coarse cloth filter.	
	Heat		Stainless steel boiling pan, heat source.	Pasteurize at 80-90°C for 10-15 min. To destroy enzymes and micro-organisms.
Fill & Seal ← Bottles/ caps		Check fill-weight and correctly sealed pack.		Bottles and caps sterilized by boiling for at least 10 minutes.
	Heat		Heat source, boiling water-bath.	Alternatively, fill juice and pasteurize in sealed bottles for 20 minutes.
Cool, Label & Store		Check that label and storage conditions are correct.		On racks, in boxes with a cool dry atmosphere.

Figure 19. - Process chart for fruit juice production

product and the bottle size, but is typically 10-20 minutes at 80-90°C. They can either be pasteurised and then hot filled into pre-sterilised bottles, or cold filled and then pasteurised in sealed bottles using a large pan of simmering water with the water level around the shoulder of the bottle. Because the pasteurisation temperature does not exceed 100°C, there is little risk of the bottles bursting. However, as with all glass containers, the temperature difference between the glass and the product or hot water should not exceed 20°C. A water cooler can be constructed to speed up the rate of cooling of filled containers. Hot bottles and cool water pass in a counter-current way through a trough to minimise 'heat shock' to the containers. In order to prevent contamination by dust, insects etc., all bottles must be cleaned properly using hand-held bottle brushes or mechanised brush cleaners. If bottles are re-used, they must be thoroughly washed with detergent to remove any residual material that may have been stored in the bottle and then sterilised by heating to boiling in a water-bath for at least 10 minutes.

Filling

Hand filling from a jug is often too slow for the required throughput and fillers can be constructed by fitting one or more taps to the base of a stainless steel or food grade plastic bucket. More sophisticated fillers measure and control the volume of liquid filled into each bottle (Figure 47).

Waxed cartons have become popular because of their lower cost compared to bottles, together with savings in time and money for collection and washing of re-useable bottles. However, the cost of equipment to form and seal the cartons is too high for most small scale producers and cheaper alternatives including plastic pots with sealed foil lids are becoming more popular in some countries.

2.2.8 Squashes, Cordials and Syrups

Squashes and cordials

These are drinks that are diluted to taste with water and are thus used a little at a time. The container must therefore be re-closeable (Figure 20) and these products may contain a preservative, usually sodium benzoate, to prevent spoilage after opening. Squashes are made from at least 30% fruit juice mixed with sugar syrup (Figure 23). Cordials are simply crystal-clear squashes. Although food dyes are used by some processors, these are not necessary for most products. Regulations on the composition of squashes are in force in some countries (see Section 2.4.2).

Syrups

Syrups are filtered juices that are concentrated by boiling until the sugar content reaches 50-70%. The heat and high solids content preserves the syrup and it is used in place of sugar or honey. Syrups can be made from a wide range of fruits, but the most common type is made from grapes.

Figure 20. - A range of squashes from different manufacturers

Stage in Process Essential / Optional	Quality Assurance	Equipment Required	Notes
Fruit pulp ↓			See Figure 19 for production of fruit pulp.
Filter pulp → Fruit	Check for uniform pulp. Ensure cloth is regularly cleaned and sterilized by boiling.	Coarse muslin cloth. Fine filter cloth or strainer.	Filter tghrough coarse muslin cloth to obtain juice and through fine cloth or special juice filters to obtain crystal-clear juice for cordials.
Heat (optional)	Check temperature and time of heating.	Stainless steel boiling pan, heat source.	Pasteurization destroys enzymes and micro-organisms which preserves the product.
Mix ← Sugar syrup	Check syrup concentration and proportions of syrup and juice.	Refractometer (optional). Heat source, boiling water bath.	Heat at 50% solution of filtered sugar syrup to 90°C. Mix in the correct proportion with the juice.
Fill & Seal ← Bottles/ caps	Check fill-weight and correctly sealed pack.	Filter, bottle capper.	
Heat (optional)		Stainless steel pan.	Heat mixture quickly to above 70°C and hot fill into pre-sterilized bottles. Heating can be done either before bottling (at 90°C for 10 minutes before hot filling into pre-sterilized bottles) or after filling with the capped bottles submerged in a tank of hot water. Most bottled squashes or cordials should be heated to 90°C for 15-20 minutes.
Cool, Label and Store	Check that label and storage conditions are correct.		On racks, in boxes with a cool dry atmosphere.

Figure 21. - Process chart for squashes and cordials

2.2.9 Preserves (Jams, Jellies, Marmalades, Pastes, Purees and Fruit Cheeses)

Jams, jellies and marmalades

Jam is a solid gel made from fruit pulp or juice from a single fruit or from a combination of fruits. The composition is controlled by law in some countries (Section 2.4.2). In mixed fruit jams, the first named fruit should be at least 50% of the total fruit content. The sugar content is normally 68-72%, which will prevent mould growth after opening the jar.

Jellies are crystal-clear jams that are produced using filtered juice instead of fruit pulp. Marmalades are produced mainly from clear citrus juices (including lime, orange, grapefruit, lemon and orange) and have fine shreds of peel suspended in the gel. Ginger may also be used alone or in combination with the citrus fruits. The fruit content should not be less than 20% citrus fruit and the sugar content is similar to jam. The correct combination of acid, sugar and pectin is needed to achieve the required gel structure and rapid boiling is necessary to remove water quickly, to concentrate the mixture before it darkens and loses its ability to form a gel. The process chart for jams, jellies and marmalades is shown in Figure 22 and a summary of common faults in jam making is shown in Table 11.

Pastes and purees

In principle, pastes and purees can be made from any fruit or vegetable. The most common types are tomato and garlic, which are widely used in cooking. These products can be made at a small scale by carefully evaporating water to concentrate the pulp, with constant stirring to prevent darkening or localised burning. The concentration of solids in the paste is normally around 36%. The high solids content and natural acidity are sufficient to preserve the product for several days but pasteurisation in bottles or cans is needed for a longer shelf life. In some preparations, sugar, salt and vinegar are added to assist in preservation.

Fruit 'cheeses'

Fruit cheeses are fruit pulps that are boiled until they have a final sugar content of 75-85%. They set as a solid block without the need for added pectin, but they do not have the same gel structure as jams and marmalades. They can be cut into bars or cubes to eat directly or they can be used in small pieces in confectionery or baked goods (see also 'fruit leathers' Section 2.2.3)

Stage in Process: Essential	Stage in Process: Optional	Quality Assurance	Equipment Required	Notes
Fruit pulp				See Figure 19 for production of fruit pulp
Filter	→ pulp	Check clarity of juice	Coarse muslin cloth	Filter through coarse muslin cloth to obtain juice.
Filter			Fine filter cloth, juice filters.	Filter through fine cloth or special juice filters to obtain crystal-clear juice for jellies or marmalade.
	Peels → Mix → Heat	Check thickness of peels and concentration of syrup.	Boiling pan, heat source, scales (pH meter optional).	For marmalade, thinly sliced citrus peels are boiled in 60% sugar syrup for 15 minutes and stored for at least 24 hours before use. For longer storage, up to 6 months, a preservative, such as 0.8% sodium benzoate can be used (also Section 2.2.3).
Mix Mix ←	Sugar Acid Pectin	Check amounts of ingredients.	Scales, scoops	Add approximately equal weight of sugar to weight of pulp and adjust with citric acid (if required) to obtain pH 3.0-3.3. For some fruits, add pectin at approximately 2% in amounts determined by trial batches.
Heat Heat		Check temperature and time of boiling.	Stainless steel boiling pan, refractometer, heat source.	Boil quickly with constant stirring to prevent burning until solids content reaches 68-70% by refractometer.
Fill & Seal ←	Bottles/ caps	Check fill weights.	Filler, capper	
Cool & Store				On racks, in boxes with a cool dry atmosphere.
Jam **Jelly or marmalade**				

Figure 22. - Process chart for jams, jellies and marmalades

Additional processing notes

Batch preparation

A simple way of calculating the amounts of sugar and juice that should be mixed together for squashes and jams is to use the *Pearson Square* (Figure 23). If for example, a squash having 15% sugar is required, made from orange juice having a 10% sugar content, mixed with 60% sugar syrup, the Pearson Square can be used as follows: draw a square, writing the juice and syrup concentrations on the left side and the product concentration in the middle, as shown in Figure 23.

Subtract the smaller amount from the larger amount diagonally to find the quantities that should be mixed together (in the example, 45 litres of orange juice should be mixed with 5 litres of sugar syrup). Similar Pearson Squares can be drawn to find the amounts that should be used when any two components are mixed together.

Orange juice

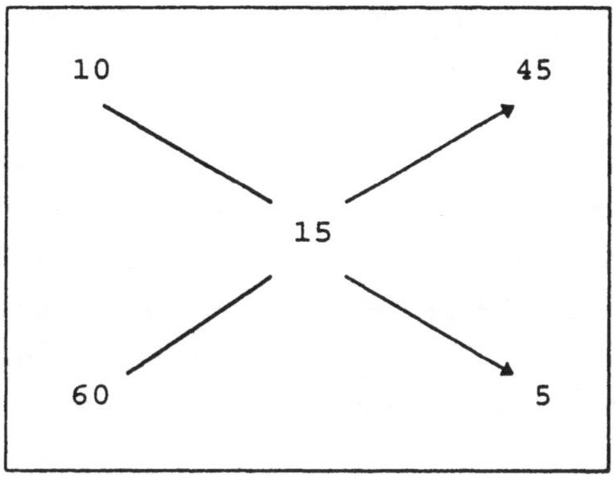

Sugar syrup

Figure 23. - Using a Pearson Square to calculate the amounts of two ingredients in a mixture. (From: *Food Chain*, N° 17, p15, March 1996)

Boiling

Boiling is done using a stainless steel pan. If other materials are used there is the risk that fruit acids will react with the pan and cause off flavours to develop. At higher production rates, a steam jacketed pan is preferable (Figure 18) as it gives more even and faster heating. If whole fruit is used, there are two heating stages in jam manufacture: initially fruit is heated slowly to soften the flesh and to extract pectin; then the mixture is boiled rapidly until the sugar content reaches 68-72%. This change in heat output requires a sufficiently large and easily controllable heat source. There are three ways to test for the correct point to stop boiling: using a hand-held refractometer (Figure 12) reading 68-70% sugar, a sugar thermometer reading 104-105 °C or by placing a drop of the product in cold water to see if it sets.

An alternative method to concentration by boiling, when making products such as tomato paste, is to hang the pulp in a sterilised cotton sack for an hour. During this time the thin watery juice leaks out and the pulp loses half its weight. Then 2.5% salt is mixed into the concentrate and it is re-hung for a further hour, during which time the weight falls to one third of the original. The product can then be packaged and pasteurised or further concentrated. This method is reported to produce a product which has a natural flavour using considerably less fuel than concentration by boiling.

Preserves should be hot filled into new or re-used glass jars, which are then sealed with a new lid. In some countries, plastic containers have been introduced but they are not easy to hot fill (they melt) and the seals are often inadequate, causing product leakage and insect infestation problems as well as a shorter shelf life.

The temperature of filling should be around 85°C. If it is too high, steam condenses on the inside of the lid and water falls onto the surface of the preserve. This dilutes the sugar at the surface and results in mould growth.

If the temperature is too low, the preserve thickens and is difficult to fill into containers. Filling can be done using jugs and funnels, but for higher production rates, small hand-operated or semi-automatic piston fillers are available (Figure 47). In all cases the jars should be filled to approximately 9/10ths full, to assist in the formation of a partial vacuum in the space above the product as it cools. This can be checked using a 'headspace gauge' (Figure 57). The jars are kept upright until the gel has formed during cooling.

2.2.10 Wines, Vinegars and Spirits

Wines are produced by fermentation in which the sugars in the fruit juice/pulp plus added sugar are converted into alcohol and carbon dioxide by varieties of the yeast *Saccharomyces cerevisiae*, named 'wine yeasts'. Producers need to select one variety that works well in their process and then continue to use it to produce a consistent product. Wines are preserved by the raised levels of alcohol and their natural acidity. Almost any fruit can be used to make wine, but the most popular in many developing countries are pineapple, papaya, grape, passionfruit, banana, melon and strawberry (or strawberry flavoured wine). Typically, the alcohol content of wine is 6-12% and in 'fortified' wines, such as sherry, ginger wine etc., it is usually 15-20%. The main problems are concerned with adequately cleaned fermentation vessels, to prevent contamination by other micro-organisms that spoil the wine and adequate sedimentation or filtration to produce a crystal-clear product. A special licence to sell alcohol is needed in most countries.

Vinegar is produced by a second fermentation in which *acetic acid bacteria (Acetobacter species)* convert the alcohol in wine to produce acetic acid, (spoiled wine is often due to the unwanted presence of acetic acid bacteria). Whereas in wine production, air is excluded from the vessel by an air lock (Figure 24), in this second fermentation it is important for the bacteria to be in contact with air as much as possible. This is done traditionally by allowing the wine to trickle over a stack of wood shavings or twigs that are held in an open framework, named a 'generator' which allows free circulation of air. The wood becomes covered in the bacteria and the process either operates continuously, re-circulating the wine a number of times until the acetic acid

Fault	Possible cause	Prevention
1. Gel does not set or is not firm	Incorrect pectin type Too little pectin Solids content too low Incorrect pH value Pectin not fully dissolved Boiling for too long Pectin solution too old Pre-setting Holding at high temperature for too long (pan too big)	Select correct type of pectin Check formulation Add more sugar Check pH, adjust with citric acid Mix with sugar before dissolving Produce smaller batches Use new stock Increase filling temperature Lower filling temperature, make smaller batches or use slow-setting pectin.
2. Gel too firm	Too much pectin Solids content too high pH too low	Check formulation Heat less, add less sugar or add more water Adjust pH
3. Pre-setting	Filling temperature too low Filling time too long Solids content too high pH too low	Increase filling temperature or choose slow-setting pectin Produce smaller batches or use slow-setting pectin See above See above
4. Fruit floats	Filling temperature too high Pectin sets too slowly Gel not strong enough Solids content too low, giving slow setting pH too high, giveing slow setting	Lower filling temperature Choose rapid-set pectin See (1) above See above Adjust pH
5. Syneresis (cracked gel with oozing liquid)	Pre-setting due to low filling temperature pH too low solids content in fruit and in gel are different	See (3) above Adjust pH Pre-mix fruit and sugar syrup and hold overnight or cook longer

Table 11. - Common faults in making preserves

levels rise sufficiently, or part of a batch is retained to inoculate the next batch. Typically, vinegar contains 6-10% acetic acid, which preserves the product for many months/years provided that it is sealed in an airtight container to prevent the acetic acid from evaporating. More modern vinegar fermenters pump air into the wine and are sufficiently expensive to be beyond the reach of most small scale producers.

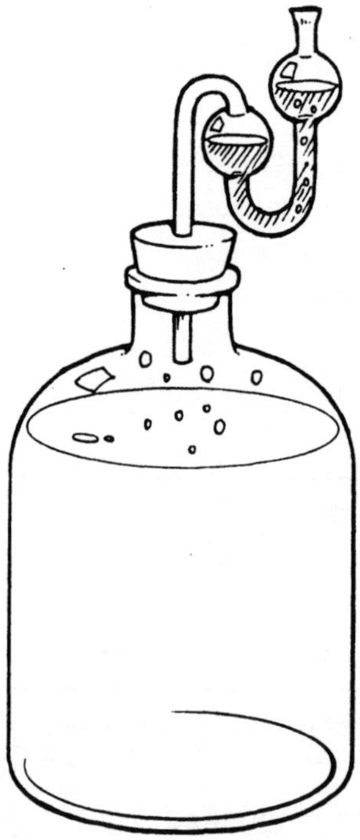

Figure 24. - An air lock used in wine making

Alcohol has a lower boiling point than water and distillation (vaporising the alcohol and then condensing it) is used to concentrate the alcohol in spirit drinks. Distillation is carried out in a still which is a clean drum, fitted with a safety valve and a pipe to carry away the vapour. Wine or other alcoholic liquor is placed inside the drum and heated. The alcohol vapour is passed through cooled air or cool water and the distillate condenses and is collected. Note: distillation is illegal in many countries without a government licence.

There are a large number of traditional methods for distillation. One such method in East Africa involves filling fruit wine or beer into an oil drum and placing it at a shallow horizontal angle over a fire (this angle prevents any liquor from being sucked over with the distillate during boiling). A hollow bamboo pipe, approximately 10cm in diameter, is fitted to the opening of the drum and sealed with a pulp made from the inner layers of a banana trunk. Five hollow bamboo pipes, approximately 2cm in diameter, are fitted vertically down from the larger pipe and sealed in a similar way. Each pipe leads to a baked clay pot of approximately 20 litres capacity, submerged in a pit of slow-running water, diverted from a stream. The pots are held below the surface of the water by wood branches that are bent over them and fixed into the soil at each end (Figure 25).

Distillation is controlled by sampling each pot at intervals to determine its alcohol strength and adjusting the rate of heating by adding or removing firewood. As the strength declines with continued distillation, the end point is determined by the desired alcohol strength in the product, usually 35-40%. Distillation is continued for 3-4 hours and 40 litres of spirit are filled from the pots into clean bottles and sealed with a stopper made from banana or papyrus leaves. The spirit is preserved by the high alcohol content and has a shelf life of several years.

Figure 25. - Traditional distillation of spirits from banana juice
(Courtesy of Midway Technology)

Modern stills are made entirely of copper or stainless steel and larger ones are fitted with thermostatically controlled heaters. These are too expensive for most small scale producers, but small copper stills for making essential oils have been used for making small amounts of alcohol.

Stage in Process: Essential	Stage in Process: Optional	Quality Assurance	Equipment Required	Notes
Fruit pulp				See Figure 19 for production of fruit pulp
	Filter		Coarse muslin cloth	Filter through coarse muslin cloth to obtain juice
	Heat		Boiling pan, heat source	Pasteurization for 15 minutes at 90-100°C destroys enzymes and contaminating micro-organisms. Cool to room temperature before adding yeast.
Mix	Sugar Yeast	Check that yeast starter culture is alive	Scales, pan	Add sugar (20%) and yeast (% of weight of pulp/sugar) to juice. Use either a starter culture of yeast or liquor from a previous batch. Yeast nutrient is optional at approximately 1g per litre.
Ferment				
Filter	Yeast and Pulp	Check rate of ferment-ation, check for mould growth and insect contamination.	Fermentation vessel, alcohol hydrometer.	It is important to prevent air from entering the fermentation vessel and an air lock (Figure 24) is commonly used. The temperature should be between 20-28°C. Fermentation may last for 5-21 days depending on the temperature. It is completed when gas bubbles are no longer produced. This liquor is used to make vinegar and spirits (Figures 27 and 28).
Clear		Check that wine is clear.	Hydrometer.	Clearing agents are needed for some fruit wines (e.g. gelatine, tannin, isinglass). Haze can also be removed using pectic enzymes. Decant cleared wine to remove sediment.
Standardize		Check alcohol content.		Adjust the alcohol content to that declared on the label by blending with previous batches or adding clean water to achieve 10-15% alcohol (15-25% for 'fortified' wines).
Fill & Steal	Bottles/ caps or corks	Check fill-weight and corfrectly sealed pack.	Filler, capper	Cold fill into pre-sterlized bottles and fit sterilized corks. (Sterilization by boiling for 10 minutes or by soaking in sodium metabisulphite solution.)
Store		Check that label and storage conditions are correct.		Store on racks, in boxes with a cool dry atmosphere away from sunlight.

Figure 26. - Process chart for fruit wine

Stage in Process Essential Optional	Quality Assurance	Equipment Required	Notes
Fermented juice			See Figure 26 for production of wine
Filter (optional)		Coarse muslin cloth.	Filter through coarse musline cloth to obtain juice.
Mix ← Acetic acid bacteria			Add either a starter culture of bacteria or liquor from a previous batch
Ferment	Check rate of ferment-ation, check for mould growth and insect contamination.	Fermentation vessel or 'generator'. Filter cloth.	It is important that air is freely circulating in the fermenter and a special design is needed to ensure this.
Filter ← Bacteria & pulp			
Clear	Check that product is clear.		
Standardize	Check acetic acid content.	Titration equipment.	Adjust the acetic acid content to that declared on the label by blending with previous batches or adding clean water or acetic acid to achieve 6-10% acid.
Fill & Seal ← Bottles/caps	Check fill-weight and correctly sealed pack.		Cold fill into pre-sterlized bottles.
Heat (optional)	Time and temperature of heating.	Water-bath.	Optional pasteurization for 5-10 minutes at 60°C.
Label & Store	Check that label & storage conditions are correct.		Store on racks, in boxes with a cool dry atmosphere.

Figure 27. - Process chart for vinegar

Additional processing notes

Sugar may be added to fruit juice to raise the level to about 20°Brix and after inoculation with about 3% yeast, it is held in a fermentation vessel, made from food grade plastic or glass, for about ten days. It is important to keep the vessel closed to prevent bacteria and mould from infecting the batch. After ten days the fermenting wine is 'racked' (filtered) by passing it through a muslin or nylon straining cloth into narrow necked fermentation vessels, plugged with cotton wool or fitted with a air lock (Figure 24).

The fermentation is then continued for between three weeks to three months, depending on the temperature and the strain of yeast being used. The end of the fermentation is seen when there are no more bubbles rising to the surface.

Stage in Process		Quality Assurance	Equipment Required	Notes
Essential	Optional			
Fermented juice				See Figure 26 for production of wine.
	Filter		Coarse muslin cloth.	Filter through coarse muslin cloth to obtain juice.
Heat			Distillation equipment.	Distil the liquor into clean containers (e.e. Figure 25).
Standardize		Check alcohol content.	Alcohol hydrometer.	Adjust the alcohol content to that declared on the lable by blending with previous batches or adding clean water to achieve 30-40% alcohol.
Fill & Seal	Bottles/ caps	Check fill-weight and correctly sealed pack.		Cold fill into pre-sterlized bottles.
Label and Store		Check that label and storage conditions are correct.		Store on racks, in boxes with a cool dry atmosphere.

Figure 28. - Process chart for spirits

It is then siphoned into clean containers and allowed to clear and mature before it is siphoned into bottles and sealed with sterilised cork stoppers or roll on pilfer-proof (ROPP) screw caps. Some fruit wines, such as pineapple, are difficult to clear because of natural gums in the fruit. These can be removed by heating the juice and allowing the gums to precipitate out before the juice is fermented. This can be a particular problem when waste fruit from other processes (e.g. drying or bottling) is used for wine making. Great care should be taken to ensure that only the fermented liquor is used for distillation. Any suspect liquor or other materials should not be distilled because of the risk that they may contain methanol. This is a type of alcohol which if consumed causes blindness and in sufficient amounts, death.

The information given in this Section is intended to enable producers to select a product for manufacture. Once a potential product has been chosen, it is then necessary to determine the feasibility of operating a small enterprise to make it and this is described in the following Section.

2.3 Conducting a Feasibility Study

2.3.1 Introduction

Aspiring entrepreneurs may have an idea about the type of fruit or vegetable product that they would like to make. This can come from seeing others successfully producing a food and wanting to copy them or from talking to friends and family members about products that they think they could make. However, an idea for a business is not a sufficient reason to begin production straight away, without having thought clearly about the different aspects involved in actually running the business. Too often, people invest money in a business only to find out later that there is insufficient demand for the product or that it is not the type that customers want to buy. To reduce this risk of failure and losing money, potential producers should go through the different aspects of running their business in discussions with friends and advisers before they commit funds or try to obtain a loan. This process is known as doing a *feasibility study* and when the results are written down, the document is known as a *business plan.*

Conducting a feasibility study need not be difficult or expensive, but the most important aspects should all be taken into account to ensure that potential problems are addressed. These are summarised in the Feasibility Study Checklist in Appendix III and are described in more detail in other Sections of this book. In this Section, the following questions that can be answered by a feasibility study are addressed:

- is there a demand for the produce? (Find out the characteristics required of the product and the size and value of the market)

- who else is producing similar products? (Determine the number and type of competitors)

- what is needed to make the product? (Find the availability and cost of staff, equipment, services, raw materials, ingredients and packaging)

- what is the cost of producing a product? (Calculate the capital costs of getting started and the operating costs of production)

- what is the likely profit? (Calculate the difference between the expected income from sales to an estimated share of the market and the costs of production)

Each of these aspects should be looked at in turn. When all the information has been gathered and analysed, it should be possible to make a decision on whether the proposed investment in the business is worthwhile or whether the producer's money could be better spent doing something else. The same considerations should be taken into account when an existing entrepreneur wishes to diversify production or make a new product.

It is also important to remember that the business plan is a working document that should be used as a framework to guide the development of a business. To do this it should be regularly updated. However, it often happens that an entrepreneur pays an adviser or consultant to prepare a business plan but then does not understand the contents, or having read it once, puts it away on a shelf never to be seen again. In the following parts of this Section, the above aspects are described in a systematic way, as should be done in a feasibility study, starting with 'The Market'.

2.3.2 Market Analysis

Once a potential producer decides that he wishes to start a business, the first thing to do is to find out what is the likely demand for the fruit or vegetable product that he or she wishes to make, by conducting a short *market survey*. Although there are market research agencies that are able to do this type of work in many developing countries, it is better for producers to do it themselves (if necessary with assistance from partners or advisers) because they will then properly understand their customers' needs and how their business should operate. If an idea is found to be feasible, this knowledge will in turn give them the confidence to go ahead when problems are encountered, knowing that their product is in demand.

Although telephone or posted questionnaires are possible, in most developing countries it is better to conduct a market survey by going out into areas where the producer expects to find consumers and asking people for their views. There are two types of information that are needed:

1) information about the product and its quality and
2) information about how much people will buy, how often and for what price.

It is important to think in advance about the type of information that is needed and to ask people the same questions each time, so that their answers can be compared and summarised. This should be a short exercise to keep the costs low and in-depth market research is not necessary for most products. A convenient way of doing this is to prepare simple questionnaires such as those shown below, which can be used by entrepreneurs to remind themselves of which questions to ask.

Product quality survey

Consumers are familiar with the types of fruit or vegetable products that are already on sale and surveys on these products are therefore easier than those for a completely new food. Questions can focus on what are the things that consumers like or dislike about existing competitors' products. For example in Figure 29 the questionnaire is used to ask questions about the qualities of chilli sauce.

However, if producers wish to make products that are new to an area, they need to have samples for potential consumers to taste and give their opinion on whether they like the product and would be willing to buy it. (When asking people to taste a product, a supply of spoons should be taken so that each person interviewed uses a clean one). Samples can usually be made at home using domestic equipment so that an investment in production facilities is not needed at the this stage. An example of a questionnaire for a new product is shown in Figure 30.

Although initially, new products have the advantage that there will be no competitors, the process of assessing demand takes longer and costs more than for products that are already known. In addition, as up to 80% of new products fail, the risks are higher and it may be more difficult to get a loan for this type of work.

Questions	1 Very good	2 Good	3 Averag e	4 Bad	5 Very bad
1. Which make(s) of sauce do you by most often?	*Write names of sauce(s)*				
1. What do you think about the colour of the sauce you buy? 2. What do you think about the seeds being present in the sauce? 3. Do you like the thickness of the sauce? 4. What do you think about the flavour of the sauce? 5. Do you like the bottle? 6. What do you think about the label? 7. What do you think about the price of the sauce?	*Tick in*	*the*	*approp*	*-riate*	*place*
8. Is there anything else that you think is good about the sauce that you buy at present?	*Write answers*				
9. Is there anything else about the sauce that you buy that you would like to see improved?	*Write answers*				

Figure 29. - Example of a survey questionnaire on the quality of competitors' chilli sauce

Remember that the customer (the person who buys the food) is not always the same person as the consumer (the one who eats the food). This is particularly important when getting information about the quality of foods that are mostly eaten by children, as their preferences for colour or sweetness may be different to those of their parents (see also Section 2.8.1). For food producers, customers can also be retailers or other sellers in addition to institutions, other food processors and members of the public.

The results of such surveys can be analysed by adding together the numbers of people that gave answers such as 'very good', 'poor' etc. In the example below, the answers to questions about chilli sauce (Table 12) show that among other information, 88% of people found the colour of the sauce to be better than average, 78% did not like having seeds in the sauce and 60% found the flavour to be good or very good.

Other information that can be gained by analysing the data includes:

1. a large majority of consumers liked having sauce in a bottle and that they were happy with existing labels. This information helps to show a new producer what type of packaging must be used if he/she is to compete effectively with existing manufacturers or imported foods.

2. A majority of consumers (52%) were unhappy with the price of the sauce and this indicates that a potential market share exists, if a new product having a similar quality can be sold more cheaply.

<table>
<tr><td colspan="6">Explain to each person you interview that you wish to start a new business making tomato jam and that you have prepared some samples for people to try. Ask them if they would like to taste the sample and give you thier opinion on what it is like.</td></tr>
<tr><th>Question</th><th colspan="5">Answers</th></tr>
<tr><td>1. Do you eat other types of Jam?
2. Which types of jam do you like best?
3. Do you think you would like tomato jam?</td><td colspan="5">Yes/No............................Circule answer
List the types
Yes/No /Not sure............Circule answer</td></tr>
<tr><td></td><th>1
Very good</th><th>2
Good</th><th>3
Average</th><th>4
Bad</th><th>5
Very Bad</th></tr>
<tr><td>4. What do you think about the colour of this tomato jam?
5. Do you like having the seeds in the jam?
6. What do you think about the flavour of this jam?
7. Do you like the texture of the jam?
8 What do you think about the jar?
9. What do you think about the label?</td><td>Tick in</td><td>the</td><td>approp-</td><td>riate</td><td>place</td></tr>
<tr><td>10. What else do you like about this jam?</td><td colspan="5">Write answers</td></tr>
<tr><td>11. Is there anything that I can do to improve this jam?</td><td colspan="5">Write answers</td></tr>
</table>

Figure 30. - Sample questionnaire for a new product (tomato jam)

Survey of market size and value

A different set of questions are needed when assessing the *size* of the market for a particular type of food (the total weight of product that is bought per month or per year) and the *value* of the market (the amount of money spent on that product each month or year). At the same time it is possible to gather information about the types of people who buy a particular food and where they buy it. A sample questionnaire is shown in Figure 31.

The information gathered from potential consumers, using questionnaires like the ones in Figures 29 to 31, can be analysed by the entrepreneur to get a good idea of the quality characteristics of the product that consumers prefer, the total demand for the product and the total value of the market. However, this involves making a number of assumptions and it is important to consider the following:

1) are the people interviewed really representative of all potential consumers?
2) were enough people interviewed?
3) were people giving accurate information?

Question	Summary of 50 replies					
	1 Very good	2 Good	3 Average	4 Bad	5 Very bad	Total
1. What do you think about the colour of the sauce you buy?	12	32	5	1	0	50
2. What do you think about the seeds being present in the sauce?	5	6	16	14	9	50
3. Do you like the thickness of the sauce?	10	20	12	7	1	50
4. What do you think about the flavour of the sauce?	42	8	0	0	0	50
5. Do you like the bottle?	40	10	0	0	0	50
6. What do you think about the label?	10	11	20	9	0	50
7. What do you think about the price of the sauce?	5	7	12	25	1	50

Table 12. - Data collected about consumers' opinions of the quality of a product

If a producer is unsure about the quality of information that has been given, he or she should ask more people the same questions to check the answers obtained. Clearly, the more people that are interviewed, the more accurately does the information reflect the real situation. However, a balance has to be drawn between the time and cost of interviewing large numbers of people and the accuracy of the data that is obtained. As a guide, 50-75 interviews should result in a good idea about the market for a product in a particular area.

When analysing data collected about market size and value, it is often helpful to find official statistics about the people who are expected to be the customers for a new product. For example in Table 13, information was collected using a market survey of chutney consumption in a small Asian town and analysed together with data from the Census Office and a previous socio-economic survey about the size and wealth of the town's population. Similar information is sometimes available from Local Government offices, tax authorities and Chambers of Commerce, although it may not always be up to date.

Questions	Answers
About the market size:	
1. How often do you buy this product?	**Daily/weekly/monthly** *(Circle answer)*
2. Do you buy different amounts at different times of the year?	Yes/No *(Circle answer)*
3. When are the times that you buy the most?	*Write answer*
4. How much do you buy each time?	*Write amount in kg or No. Packs*
5. When are the times you buy the least?	*Write answer*
6. How much do you buy each time?	*Write amount in kg or No. Packs*
7. What is the amount of food in the pack?	*Write amount in kg*
About the market value:	
8. How much do you pay for a pack of the food?	*Write the amount in currency*
9. What is the price difference for larger or smaller packs?	*Write differences*
10. Does the price change at different times of the year?	Yes/No *(Circle answer)*
11. When is the price highest?	*Write answer*
12. When is the price lowest?	*Write answer*
About the customer:	
13. Would you say that you have a low, medium or high income in your household?	Low/medium/high *(Circle answer)*
14. In which age group do you belong?	*Tick answer*
1-20	
21-40	
41-60	
Male/Female	M/F (*Circle answer*)
About sales outlets:	
15. Where do you usually buy this food:-	*Tick answer*
Market stall	
Local shop	
Kiosk	
Supermarket	
Street hawker	
Directly from producer	
Other	*Write answer*

Figure 31. - Sample questionnaire about market size and value

Type of customer	Number in each category*	Amount of chutney bought per month (kg)**	Amount of chutney bought each time (kg)**
Low income	18,430	0.4	0.1
Medium income	5, 485	1.2	0.15
High income	192	2.25	0.45

* from official census statistics for the town

** average of information given by 70 customers interviewed

Table 13. - Potential market for chutney in a small Asian town

The cost of chutney in the market was $3.9 per kg when sold in 100g amounts from a bulk container into customers' own pots (bought by the majority of those who said they were in low income families), $4.1 per kg when sold in 150g plastic bags (bought mostly by medium income families) and $4.8 per kg when sold in 450g glass jars (bought mostly by high income families). This data can be analysed, as shown in Tables 14 (a) and (b), to calculate the total size and value of the chutney market in this town.

The size and value of the market, calculated in Table 14, indicate that low income and medium income families form the largest part of the chutney market in this town. These people were found to buy the product either from bulk containers into their own pots, or pre-packed in plastic bags. The demand for jars of chutney was limited to high income groups which formed only 3% of the market size and 3.8% of its value. A new business would therefore be likely to focus on low and medium income families as its potential consumers. This has implications for not only the type of packaging that is used but also the types of advertising, methods of promotion and agreements with retailers that should be considered. These aspects are described in more detail in Section 2.8.

Type of customer	Number in each category	Amount of chutney bought per month (kg)	Total demand (kg per month)
Low income	18,430	0.4	7,372
Medium income	5,485	1.2	6,582
High income	192	2.25	432
TOTAL			**14,386 kg**

Table 14 (a). - Calculation of the size of the market for chutney

Type of customer	Amount of chutney bought each time (kg)	Cost per kg of chutney ($)	Number of kg	Value of market ($ per month)
Low income	0.1	3.9	7,372	28,751
Medium income	0.15	4.1	6,582	26,986
High income	0.45	4.8	432	7,882
TOTAL				**$63,619**

Table 14 (b). - Calculation of the value of the chutney market

Market share and competition

Market surveys and the calculation of market size and value are important to find out whether the demand for a product really exists, but these figures should not be assumed to represent the scale of production that could be expected. Even if no-one else is currently making a product locally, it is likely that once a new business starts production and is seen by others to be successful, they too will start up in competition. It is therefore important from the outset, to estimate what is the proportion of the total market that a new business could reasonably expect to have. This is known as the *market share*. It is often difficult to estimate a realistic market share and the figure depends on a large number of variables, but Table 15 can be used as an initial guide. In many cases, new entrepreneurs over-estimate the share that they could expect, with the result that production operates at only a small proportion of the planned capacity. The lower percentages in Table 15 should therefore be used initially. Section 2.3.4 describes some of the negative effects on finances of operating a business below planned capacity.

In the example described in Tables 13 and 14 concerning the market for chutney, there were a large number of small producers all making similar products. The estimated share of the market for a new producer can therefore be calculated as follows:

Total size of the market = 14,386 kg per month

Estimated share = 5-10%, with 5% selected.

This represents sales of 719 kg of chutney per month with a potential value of $3,181 per month.

When converted to daily production rates, assuming 20 working days per month, the maximum production is therefore <u>36 kg per day</u>

This figure for daily production rate is very important. It is central to all subsequent calculations of production capacity and investment requirements (below) and every care should be taken to ensure that this information is as accurate as possible.

It should be noted that in the calculations below, the scale of production is based on an anticipated share of the total market. In other situations, a more detailed analysis of market segments could be made (Section 2.8.1) and the planned market share could be based on one of those segments (e.g. low income groups in Tables 13 and 14).

Competitors are very important to the success or failure of a new business and the entrepreneur should recognise that there are different types of competitor. Using the example of someone wishing to make fruit juices, it is helpful to think how the consumers might view the available products: for example when they are thirsty, they have a choice of hot drinks (tea, coffee etc.), cold soft drinks, such as milk, juices, squashes or finally alcoholic drinks. These are all *general competitors*, who are able to satisfy the consumers' thirst. Supposing the consumers choose cold soft drinks that can be drunk straight from the bottle, they then have a choice between carbonated (fizzy) soft drinks, and juices. These are known as *type competitors* or different kinds of soft drink. Finally, on choosing juices, there are different juices and different brands of the same type of juice, which are *brand competitors*.

Although the appearance and quality of foods are important, competitors do not just compete with their products. They also compete with the profit margin and level of service that they offer to retailers and with special offers or incentives to customers. New entrepreneurs must therefore assess each of these factors when deciding what the competition is and how to deal with it. This is conveniently done using a *SWOT analysis*, where SWOT stands for <u>S</u>trengths, <u>W</u>eaknesses, <u>O</u>pportunities and <u>T</u>hreats.

The technique involves looking at each aspect of the new business and comparing it to other producers, particularly type and brand competitors.

Many new entrepreneurs do not appreciate the importance of finding information about competitors and even if they do, they may not know where to find it. In addition to the direct questions to consumers in market surveys described above, entrepreneurs can get information about competitors from the following sources:

1. discuss with retailers the amount of sales of different brands and any seasonality in demand. What are the trends in consumers' buying, what is getting popular and what is going down? What types of consumers buy particular products and how often? Does the retailer put on any special displays for some suppliers? What do they think about the idea for a new product and do they think they will sell a lot of it? What are their plans for the future?

2. look at competitors' advertising and retail displays, get a copy of their price lists.

3. ask the local Employer's Federation or Chamber of Commerce for any information they have on the market for similar products.

4. visit trade fairs and talk to other producers and their customers.

5. look in trade journals, manufacturers' association magazines and newspapers for information about the market and the activities of competitors.

No. Of other producers	Many				Few				One				None
Size of competitors	Large		Small		Large		Small		Large		Small		
Product range	S	D	S	D	S	D	S	D	S	D	S	D	
Market share (%)	0-2.5	0-5	5-10	10-15	0-2.5	5-10	10-15	20-30	0-5	10-15	30-50	40-80	**100**

S = similar products , D = dissimilar products

Table 15. - Estimates of market share for a new food business with different levels of competition
(From Do Your Own Scheme, Anon)

After finding as much information as possible, the entrepreneur can then start to compare the new business with those of competitors using the SWOT analysis. An example of how it might appear is shown in Table 16.

When it is completed, the entrepreneur should be able to answer the following questions:

- who is producing similar products?
- where are the competitors located?
- what is the quality and price of their products?
- what can I do to make a new product that is better than those of competitors?
- why would customers or consumers want to change to a new product?
- what offers or incentives do competitors give to retailers?
- what are competitors likely to do if a new product is introduced?

The answers to these questions are then used to formulate a marketing strategy, details of which are described in Section 2.8.2.

The analysis in Table 16 indicates that one competitor (A) has a range of good quality products that are packaged and promoted well, but they are more expensive and do not meet changing consumer requirements. The other competitor has a cheap product that is not well packaged and not promoted. However, it sells well because the low price attracts low income consumers and retailers promote it because of the higher margins offered by the company. They appear to be expanding to new areas. However, retailers are annoyed when Competitor B fails to deliver on time or in the correct amount and they may have over-stretched their distribution capacity. The analysis points the way to producing a product without additives and to providing a good service and equivalent margins for retailers. It also highlights lack of information about process inputs (e.g. packaging) and production costs. These are discussed in the Sections below.

2.3.3 Technical Feasibility

Once an entrepreneur has found information about potential consumers, their requirements and the likely share of the market that could be obtained for a new product, it is then necessary to assess whether production at this scale is technically feasible. The series of questions below is helpful in deciding the technical requirements of the business:

	My proposed business	Competitor A	Competitor B
Strengtha	Production likely to be sited close to retailers - can deliever at short notice.	Good brand image and range of products.	Product is cheaper than A and sells well. They offer good margin to retailers.
Weaknesses	Difficult to find good packaging.	Products more expensive than B. Uses synthetic colours and preservatives.	Poor quality product, poor label design. I'm told by retailers that suppies are irregular and not always the amount ordered.
Opportunities	Retailers say demand for products without additives is increasing. I can produce without added colours.		Appears to be expanding deliveries to new areas according to newspaper reports.
Threats	Strong promotion by A. There are few wealthy consumers and price is most important factor. I am not yet sure of production costs.	Cheaper products than B.	May have over-expanded distribution network and failing to make deliveries.

Table 16. - Example of a SWOT analysis of a new business in relation to competitors

- are enough raw materials available of the correct quality when needed for year-round production?
- is the cost of the raw materials satisfactory?
- is the correct size and type of equipment available for the expected production level and at a reasonable cost?
- can it be made by local workshops and are maintenance and repair costs affordable?
- is sufficient information and expertise available to ensure that the food is consistently made at the required quality?
- are suitable packaging materials available and affordable?
- are distribution procedures to retailers or other sellers established?
- is a suitable building available and what modifications are needed?
- are services (fuel, water, electricity etc.) available and affordable?
- are trained workers available and are their salaries affordable?.

(Adapted from: *Starting a Small Food Processing Enterprise*, by Fellows, Franco and Rios)

Production planning

The answers to these questions can be found by first setting down a plan of the production process in a similar way to the process charts described in Section 2.2. This plan should indicate how the different stages in a process are linked together, identify any 'bottle-necks' in the process, the equipment that is required for each stage and where quality assurance procedures should be used. The data that has been found from market surveys is added to the process chart to indicate the scale of production that is required (e.g. Figure 32, which uses chutney as an example).

The chart is also used for planning a number of different aspects of the production process, including:

1) the weights of raw materials and ingredients that should be scheduled each day,
2) the number of workers and their different jobs,
3) the size of equipment required to achieve the planned throughput of product
4) the number of packages that are required each day.

In the example, the market information for chutney sales indicated that a minimum production rate of 36 kg per day would be needed to meet the anticipated initial market share. Assuming that production takes place for 8 hours each day for 20 days per month, the average throughput would be 4.5 kg per hour (36/8kg). This throughput figure is critically important in all subsequent planning and every effort should be made to ensure that it is as accurate as possible by checking all assumptions carefully. In particular, the number of assumed working days may fall below twenty if there are regular power failures or if production planning (Section 2.7.1) is inadequate. The different stages of production planning are described below.

Processing stage	% losses	Weight of mangoes (kg)	Batch size (kg)	Processing time (minutes) from Figure 33.	No. of workers	Minimum equipment size (kg/hr)
Mangoes	0	60				
Wash	14	60				
Sort	45	51.6		90	2	Table for 2 workers
Peel/destone	0	28.4				
Cut		28.4		120	3	Table for 3 workers, 3 knives
Mix	0	27	27 kg sugar + 13.51 vinegar for batch of 60.7 kg.			
Boil	34*		40	180	1	Boiling pan for 10 kg batches. Two filters and heat sealers. Table for 2 people
Fill/seal	10		36	180	2	
Cool/label	0		36	120	1	
Store	0		36			
Weight of product			36			

* evaporation losses during boiling. Note: recipe described in Figure 13.

Figure 32. - Modified process chart showing scale of operation and daily requirements for mango chutney production

Notes on calculations:

Boiling results in weight losses of 34% as water is evaporated and the solids content increases to 70% (see calculation below). If each batch takes 20 minutes to boil, there are 2 batches per hour and in 3 hours there are 6 batches of 10kg each to meet production target of 60 kg of raw materials, yielding 36 kg of product per day. Therefore the boiling pan should have 10kg working capacity (that is a 12-15 litre pan).

Each worker fills and seals 40 bags per hour = 120 bags per day x 2 workers
= 240 bags of 150g net weight
= 36kg per day

Calculation of boiling losses:
The solids content in the mix of ingredients before boiling is found as follows:

Ingredient	Weight (kg)	Solids content (%)	Weight of solids (kg)
Mangoes	27	15	4.05
Sugar	27	100	27
Vinegar	13.5	0	0
Total	67.5		31.05
Total weight after 10% losses	60.7kg		28

% solids in batch before boiling = (28/60.7) x 100
= 46%

So 28kg equals 46% of the batch before boiling. After boiling there is no loss of solids (only water is removed) but the solids content has been increased to 70%.

Therefore 70% still equals 28kg.
Therefore the total weight of the batch after boiling = (100/70) x 28
= 40kg

Weights of raw materials and ingredients

There are two stages involved in planning the amounts of materials that are needed to produce the required weight of product: first, it is necessary to calculate the amount of each ingredient that will be needed to formulate a batch of product and secondly, it is necessary to calculate the amount of losses that can be expected during preparation of fruits and vegetables.

The processor should experiment with different mixes of ingredients (the 'formulation' or 'recipe') to produce a product that has the colour, flavour, appearance etc. that consumers say they prefer from market research. Skill and flair are needed to achieve this, using the combination of ingredients that has the lowest cost. It is important to weigh each ingredient carefully and make sure that all weights are recorded for each formulation that is tried.

Otherwise, the inevitable result is a successful trial product, but no information is recorded to enable it to be repeated. Once a formulation has been successfully developed, great care is needed to ensure that it is made in exactly the same way on every occasion. This requires staff training, especially for those staff involved in batch preparation, the implementation of quality assurance procedures and careful production control. These aspects are discussed in more detail in Sections 2.7.1 and 2.7.2.

Nearly all fruit or vegetable processing results in losses of material. These may arise from peeling or de-stoning, from unsatisfactory fruits and vegetables that are thrown away during sorting, from spillage during filling into packs or from food that sticks to equipment and is lost during washing. Different types of fruit and vegetables have been found in practice to have different levels of wastage and examples of some of these are given in Table 17. Typical losses from other sources in a well-managed production process are shown in Table 18. However, it is necessary for an entrepreneur to do trials to calculate the actual amount of wastage experienced with the particular varieties of fruit or vegetable and with the particular process that are being used.

Clearly, it is in the interests of the processor to reduce losses as much as possible. Contracts with reliable suppliers (Section 2.6.1) help to ensure lower levels of poor quality raw materials and therefore reduce losses. Additionally, a well-managed processing operation, having good quality assurance procedures (Section 2.7.2), also reduces wastage, especially during later stages of a process when the product has a higher added value.

Using the data from experimental production trials, or less desirably estimates based on data in Tables 17 and 18, it is necessary to calculate the amount of raw materials and ingredients that are needed to produce the required weight of product each day. This will also enable the true cost of raw materials to be calculated for use in financial planning (Section 2.3.4)

Using mango chutney as an example, Figure 32 shows losses during each stage of the process. The amount of mangoes that need to be bought to produce the required weight for each day's production can then be calculated. The result indicates that only 45% of incoming raw materials were actually used in the product (27kg of the 60kg bought). If mangoes were bought for $0.2 per kg in season, the true cost of the fruit is calculated as $0.44 using the following formula:

$$\text{True raw material cost} = \frac{\text{Supplier cost}}{\text{\% Yield}} \times 100$$

$$= (0.2/45) \times 100$$

$$= \$\ 0.44/\text{kg}$$

Other ingredient costs are estimated as follows: sugar $0.6/kg, vinegar $1.25 per litre and total spice costs of $1.3 per day. This data is used to calculate operating costs in Section 2.3.4.

Fruit or vegetable	Typical losses during preparation (%)	Notes
Apples	23	peeled & cored
Apricot halves	12	destoned
Bananas	41	peeled
Cabbages	30	
Carrots	4	(bought without leaves)
Cauliflowers	38	
Currants	3	seeds & skins removed
Figs	2	
Grapes	19	skins & pips removed
Guava	22	
Lemons	40	peel & seeds removed
Mangoes	45	peeled & destoned
Melons	42	peel & seeds removed
Okra	12	
Onions	3	
Oranges	25	peel & seeds removed
Passion fruits	58	peel & seeds removed
Pawpaws	38	peel & seeds removed
Peas	50	bought in pods
Peppers - chilli	15	seeds & stalk removed
Peppers - green	14	seeds & stalk removed
Pineapples	48	peeled & cored
Plantains	39	peeled
Tomatoes	4	seeds & skin removed

Table 17. - Typical losses during the preparation of selected fruits and vegetables (Adapted from data in *The Composition of Foods* by Paul and Southgate, and from field data collected by the author)

Equipment required

Using the process chart (Figure 32), the weight of food that should be processed at each stage is then calculated in kg per hour. This information then allows the processor to decide what equipment is required and the size (or 'scale' or 'throughput') that is needed. In doing this, decisions need to be taken on the relative benefits of employing a larger number of workers or buying machinery to do a particular job. In some enterprise development programmes, there may be wider social objectives of employment creation which may influence such decisions.

The decisions on equipment requirements are also influenced by:

- the cost and availability of machinery
- the availability of people who are skilled in
- maintenance and repair
- the availability and cost of spare parts and
- the possibilities of local equipment fabrication.

Information on the types and suppliers of equipment is often difficult to obtain, but catalogues and sometimes databases of equipment manufacturers and importers may be available at offices of national and international development agencies, Chambers of Commerce, university departments, food research institutes, embassies of other countries and trade or manufacturing associations.

Stages in a Process	Typical losses
Washing fruits/vegetables	0-10
Sorting	5-50*
Peeling	5-60
slicing/dicing	5-10
Batch preparation/weighing	2-5
Boiling**	5-10
Drying**	10-20
Packaging	5-10
Machine washing	5-20
Accidental spillage	5-10
Rejected packs	2-5

* Unsatisfactory raw materials depend on source and agreements with suppliers
** does not include evaporation losses

Table 18. - Typical Losses During Processing of Fruits and Vegetables.

It is preferable wherever possible, to buy equipment from local suppliers and fabricators because servicing and obtaining spare parts should be faster and easier. However, if equipment has to be imported, the following points should be considered:

when ordering equipment, it is important to specify exactly what is required, as many manufacturers have a range of similar products. As a minimum, it is necessary to state the throughput required in kg per hour and the type of food to be processed. Where possible other information such as the model number of a machine, whether single or three-phase power is available and the number and types of spares required, should also be given. Assistance from a food technologist working in a local university or food research institute may be required to research and order equipment. The quotations received from equipment suppliers can then be used when calculating financial viability (below).

Packaging

Similar considerations apply when ordering packaging materials as there is a very wide range available and there are a number of considerations that should be taken into account by the producer. These include the technical requirements of the product for protection against light, crushing, air, moisture etc. (described in Section 2.5.5. and for individual products in Section 2.2), the promotional and marketing requirements (Section 2.8.3) and the relative cost and availability of different types of packaging. Selection of packaging materials frequently causes the largest problems for small producers and is often the main cause of delay in getting a business established. Professional advice should be sought from a food technologist or in some countries, packaging specialists or agents of packaging manufacturers.

Staffing levels

Decisions on the numbers and types of workers that are required to operate the proposed business are taken in conjunction with decisions on equipment procurement. Using the process chart, it is possible to break down the production into different stages and then decide the number of people who will be needed for each stage of the process. It is important also to include work such as store management, quality assurance and book-keeping when planning employment levels.

In fruit and vegetable processing, each day's work will initially involve preparation of the raw materials and then move through processing to packaging. It is possible to have all workers doing the same type of activity throughout the day, but it is often more efficient to allocate different jobs to each worker as the day progresses. A convenient way of planning this is to draw an *Activity Chart*. This shows the type of work that is to be done each hour during the day, the number of people involved with each activity and the sequence of work that individuals will do during the day.

In the example of chutney processing, the total number of workers is estimated from the process requirements shown on the process chart (Figure 32). It is estimated that two workers will be able to wash and sort 40kg mangoes within ninety minutes. Similarly, it will require three workers to peel and slice this amount of fruit within two hours (Figure 33). Once sliced fruit becomes available (by around 9.30 am), one of the three workers (X) can begin preparing the batches of ingredients and boiling the chutney. By 11.00 am, fruit preparation has finished and while one worker (Y) washes down the preparation area, the third (Z) labels the previous day's production and packs them into boxes ready for distribution.

In this plan, all workers have a lunch break at the same time, but in other types of process it may be more convenient or efficient to stagger each person's break at different times. As the first batch of product cools sufficiently, work can begin after lunch on filling and sealing it into 150g plastic bags. This is a time-consuming stage as manual filling and sealing have been selected. Additionally, packages require check-weighing to ensure that they contain the correct weight of product (Sections 2.4.2 and 2.7.2). It is calculated that three hours will be needed for two people to fill and seal 240 bags (36kg). This time could be reduced if a mechanical filler/sealer was bought, particularly at a later time when the business expanded. In the example, the owner/manager (M) is involved with staff supervision, record keeping, finance management and product distribution/sales. In other plans, these jobs could be done by trained staff.

	Hours during working day										
Activity	8 am	9am	10am	11am	12pm		1pm	2pm	3pm	4pm	5pm
Washing/ sorting	X Y				L						
Peeling/ slicing		X Y Z			U						
Mixing/ boiling			X		N						
Cleaning				Y				Z			
Packing					C		Y Z				
Labelling	Z			Z							
Store manage- ment					H				X		
Distrib- ution										M	
Manage- ment	M										

Figure 33. - Activity chart used to plan job allocations for staff to produce mango chutney

This type of chart is useful for assessing the time required to complete each stage of the process and for thinking through the problems that are likely to occur. When production begins, it can be used as a basis for training in each job and it should be constantly reviewed to optimise production efficiency.

In summary, the technical part of a feasibility study involves taking information about the expected demand from the market survey and calculating the process throughput required to meet that demand. This can then be used to decide on the type of equipment, the level of staffing and the amounts of raw materials, ingredients and packaging that will be required. These are summarised, using the example of chutney production, in Table 19.

Information required	Data obtained
Estimated market size (kg/month)	14,386
Estimated share of market	5
Production required per month to meet market share (kg)	719
Production required per day @ 20 days' work per month (kg)	36
Minimum Process throughput @ 8 hours per day (kg/hr)	4.5
Weight of mangoes required per day (kg)	60
Losses on arrival due to sorting (%)	14
Amount of losses in the process (%)	
- wastage/spillage	10
- peeling losses	45
- mixing losses	15
- packing losses	10
- evaporation losses during boiling (%)	34
Minimum size of equipment required (kg/hr) for	
washing/sortin	60
peeling/slicing	40
boiling (2 batches of 10kg per hour)	10
packing (bags per person per hour)	40
Number of people required to operate the process	3 plus owner/manager

Table 19. - Summary of technical feasibility calculations for mango chutney production

2.3.4 Financial Feasibility

Having completed the study of technical feasibility, the entrepreneur should then have sufficient information to determine the costs that are likely to be involved in production. Additionally, the market survey will have supplied information about the sale price that could be achieved for the new product. The entrepreneur is therefore in a position to calculate the expected income and expenditure and hence the gross profit that can be achieved.

Start-up costs

When a new fruit and vegetable processing business is started, it is likely that money will be required to buy or convert a building and buy equipment to start production. Details of suitable buildings are given in Section 2.5.3. Additionally, it is necessary to buy a stock of packaging materials and the initial raw materials and ingredients. The *start-up capital* is the amount of money that is needed to buy the facilities and equipment, to register and licence the business and get the necessary hygiene certificates.

Working Capital includes the costs of raw materials, packaging, staff training, product promotion etc. that have to be made before the business begins to generate income from sales of the product. The requirement for working capital also continues as the business develops and is discussed further under 'Cashflow' below. As described in Section 2.7.1, fruit and vegetable processing has relatively high requirements for working capital compared to other types of food processing. This is because of the seasonal nature of crop production and the need to buy several month's supply of crops during the season and part process them so that production can continue for a larger part of the year.

The start-up capital and initial working capital are calculated to determine whether the entrepreneur's savings (known as the *owner's equity*) will be sufficient to start the business without a loan. Using the example of chutney production, the start-up costs are estimated in Table 20, using representative data from the country concerned.

Start-up cost	$
Conversion of building (Section 2.5.3)	800
Equipment (from Figure 32)	350
Registration of business	50
Business Licence	25
Hygiene insepction and certificate	50
Raw materials & ingredients for 4 weeks' production (from Figure 32)*	927.5
Packaging (minimum order)	200
Staff training (equivalent to income from 2 weeks' production value)**	1476
Initial production promotion	250
Staff salaries for 6 weeks	360
TOTAL	**4488.5**

* 60kg mangoes/day @ $0.2/kg = $240/month,
27kg sugar/day @ $0.6/kg = $324/month,
13.5 litres vinegar/day @ $2.25/litre = $337.5/month,
Spices cost $1.3/day = $26/month

** Sales @ $4.1/kg (Table 14) x 36kg/day = $1476 for 2 weeks.

Table 20. - Start-up costs for chutney production

The owner's equity is $2,500 and a loan of $1,989 is taken to meet the total startup costs. (A further option of a second partner's equity of $2,000 is agreed at the same time to take account of a negative cashflow during the first year of operation (see Table 22)).

Operating costs

There are two types of operating (or production) costs: those expenses that have to be paid even if no production takes place and those that depend on the amount of food that is produced. The first type are known as *fixed costs* and the second type are *variable costs*. Examples of each are shown in Table 21 again using chutney production as an example.

Type of Production Costs	Actual Costs for Chutney Production per Year ($)
FIXED COSTS	
Rent	1200
Labour*	2880
Loan repayment**	19898
Interest charges**	796
Professional fees (e.g. accountant's fees)	120
Maintenance of equipment (10% of value)	35
Depreciation of equipment (over 3 years)	117
Business registration fees, hygiene certificates and other licences	125
Total fixed costs	**7262**
VARIABLE COSTS	
Raw materials (Table 20)	2880
Other ingredients	8250
Fuel	800
Power	250
Packaging materials	1800
Transport/distribution	450
Labour*	-
Advertising and promotion	1150
Total variable costs	**15,580**
Total operating costs per Year	**22,842**

* Labour is a fext cost if workers are permanently employed as full-time staff, but it is described as a variable cost if people are only employed when production takes place. In this example, permanent labourers are paid $80/month.

** In this example, the loan of $1989 is repaid within the first year with a fixed interest rate of 40% per month.

Table 21. - Summary of fixed and variable operating costs for mango chutney

Income and profit

From the market survey, the estimated market size and share enables the expected sales to be calculated. The *gross profit* (or gross loss) is the difference between the expected income and the total operating costs over the first year, including any loan repayments. Income is therefore calculated as follows:

Income = Selling price per unit x number of units sold

The income clearly depends on both the price of a product and the amount that is sold. When selecting a price for a product, two approaches can be taken: first the price can be based on production costs and it is set to ensure that income exceeds the total costs. This however, does not take account of competitors' prices and to be successful, the new product should be priced at or below the price of other similar products.

The second approach is therefore to set the price to compare favourably with existing products and calculate the likely profit at the planned scale of production.

Unless the new product is to be sold directly from the production unit or through a sales outlet owned by the producer, it is also important to remember the profit that will be expected by retailers. In many countries this profit is normally 10-25% of the value of each pack. In addition, there are distribution costs and perhaps special promotion costs that should also be included. The price that is charged for the product should therefore allow the producer, the distributors and the retailers to make an adequate profit. In the example using mango chutney, the income to the producer is the sale price less 10% for retailer's profits ($4.1 - 10% = $3.7/kg).

When the production costs and income are compared using the second approach, the operation of the business should be above the b*reakeven Point*. Above this point is the minimum level of production that can enable the enterprise to make a profit (Figure 34).

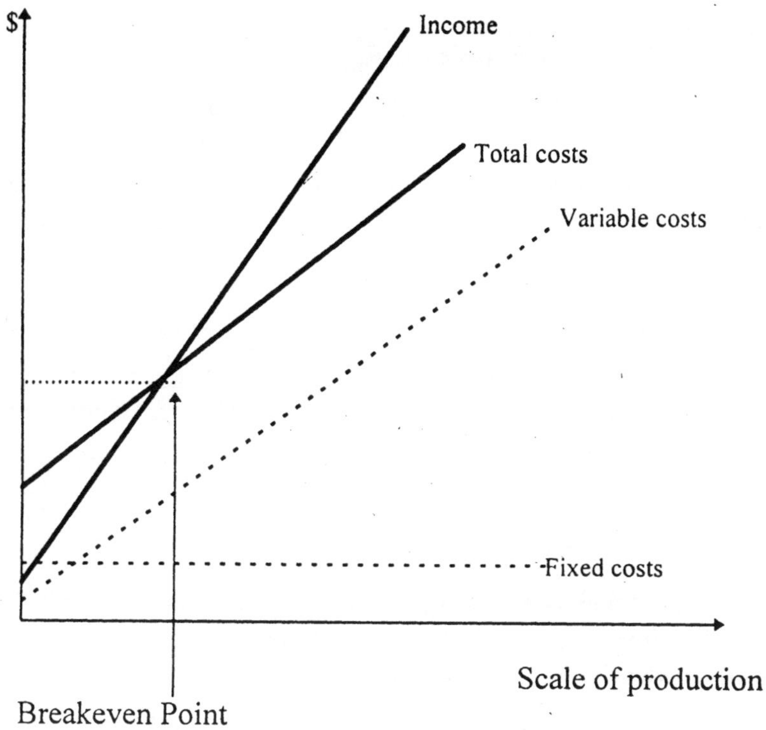

Figure 34. - Breakeven Point

Breakeven point can be calculated as follows:

- calculate the contribution for variable costs per pack
- subtract the value obtained from the sale price to obtain the 'unit contribution'
- calculate the total fixed costs per year
- divide the fixed costs by the unit contribution to obtain the annual production rate that will allow the business to break even

In the example of chutney production, the contribution for variable costs per pack (Table 21) = $ 15,380/57,600 bags per year = $ 0.270

The sale price per pack = $ 3.7/kg /6.67 packs per kg
= $ 0.555 per pack

Unit contribution = sale price - (variable + labour contributions)
= 0.555 - 0.270
= 0.285

Total fixed costs per year = $ 7262

Breakeven = fixed costs/unit contribution
= 7262/0.285
= 25,481 packs per year

When expressed as a % of total production capacity (57,600 bags per year),
the breakeven point = (25,48157,600) x 100
= 44.2%

In other words, the processor must operate at above 44% of the available capacity in order to make a profit. Clearly the higher the figure for the break-even point, the more difficult it is for a process to be profitable.

The annual production costs are calculated in Table 21 as $22,842. If all products are sold, the annual income is calculated to be $31,968 (36kg per day @ $3.7/kg x 240 days per year). This leaves a gross profit of $9,126 per year, which after taxes, is available to pay the owner a salary and for re-investment and expansion of the business.

If the feasibility study shows that the scale of production required to meet the expected market share is below the break-even point, the entrepreneur should carefully examine the data to see if production costs can be reduced. If not, there is a question over the wisdom of proceeding further with the proposed business.

It should be noted that entrepreneurs should not automatically consider the gross profit as their own income. The money belongs to the business and they should take a fixed wage, which is recorded as another business expense. A common source of business failure happens when an owner removes cash to pay for a funeral or other family occasions and disrupts the cashflow of the business to a point that it cannot continue trading.

Financial Planning

If the gross profit indicates that the proposed fruit and vegetable processing business is likely to be successful, it is then necessary to repeat the calculation of monthly gross profit for one to three years. This will then show whether there is sufficient cash available to operate the business without the need for further loans. This is known as a *cashflow forecast* and an example, calculated for one year only for chutney manufacture, is given in Table 22.

Month	J	F	M	A	M	J	J	A	S	O	N	D	Total
Income ($'000)	0.4	0.6	1.0	1.2	1.2 + 2.0	1.9	2.1	2.2	2.5	2.7	2.7	2.7	22.2
Expenses ($'000)	1.0	1.2	1.4	1.5	1.7	1.8	1.8	1.8	1.8	1.8	1.8	1.8	19.4
Cumulative Profit/loss ($'000)	(0.6)	(1.2)	(1.6)	(1.9)	(0.4)	(0.3)	0	0.4	1.1	2.0	2.9	3.9	2.8

Figures in ( ) indicate a negative cashflow. The second partner's equity of $2000 was taken in May.

Table 22. - Example of cashflow forecast for chutney manufacture

From the data in Table 22 it can be seen that during the initial startup period during January and February, production routines were becoming established and as a result, sales were low. The expenditure on supplies of packaging materials and fruit during this time leads to an accumulated negative cashflow of $1,900 by April. This illustrates one of the benefits of conducting a feasibility study: the losses made over the first few months are planned and can be addressed by taking out a loan or using the owner's equity. This gives both the owner and any lenders the confidence to know that the business is under control and that the negative cashflow will cease, in this case after seven months. Lenders are more willing to provide a loan if they are confident that the finances of the business are planned and managed. This should not be done just at the start of a business but also later on, if sales are expected to fall for a while or if raw material costs rise temporarily (e.g. when the harvest season finishes). A particular problem for all small businesses is the need to order packaging materials in bulk because of minimum order sizes. This expenditure and the need to tie up cash in stored packaging can be very damaging to a business cashflow. The entrepreneur should assess the alternatives of paying a higher unit price for small amounts of packaging or suffering a negative cashflow.

A similar forecast is made to show the expected development of the business over three years (not forgetting to take account of the expected actions of competitors). Finally, in assessing financial feasibility, the data is presented as a *Profit and Loss Statement*, to calculate the net monthly profit before tax over the first three years. An example of a monthly profit and loss account is shown in Figure 54.

Preparing a business plan

The advantages of writing down the results of the feasibility study are as follows:

- the findings can be set out in a clear and logical way, so that potential lenders can understand the business and its likely risks/advantages

- the document helps the entrepreneur to clarify and focus his/her ideas

- it is reference material that can be used to plan long term development of the business

- the plan can be regularly consulted and updated as a guide to the business development

- mistakes can be made on paper rather than in the operation of the business

- when the plan shows that a successful business is possible, it makes the entrepreneur feel more confident about success

- it helps the entrepreneur to decide how much money is needed and if properly prepared, it gives the loan agency confidence that their money will be repaid.

Most lenders have little understanding of fruit and vegetable processing and the entrepreneur should therefore write the business plan in a simple way, avoiding jargon and technical language as much as possible. If lenders can understand what is involved in the business, they are more likely to approve a loan.

It is important to include as much detail as possible and if necessary do thorough research first. It is also important to look outwards from the business to judge what competitors will do and how the business will develop to become sustainable.

Although there is no fixed way of writing a plan, the sections that could be included are summarised as follows and in Appendix III:

Introduction:	to summarise what the product is, who is expected to buy it, why the business is a good idea,
Basic information:	the name and address of the owners, their qualifications and experience,
The product:	details of the raw materials, the production process, quality assurance, packaging etc. What is special about the product compared to those of competitors,

The market: the potential customers, where they are located, the size and value of the market, expected market share, likely expansion (or contraction) of the market, the number and types of competitors, their strengths and weaknesses and their expected reactions to a new product,

Selling plan: distribution and sales methods, planned promotion, product cost,

Premises/equipment: where the business will be located, building to be used and services that are needed, steps taken to meet health and hygiene laws, equipment and its cost,

Finance: amount required for start-up and initial operation, including profit and loss statement and cashflow forecast for three years, owner's resources that will be used, size of loan required and what it is for, security on the loan,

Business registration: steps that have been taken or are planned to register the business with tax authorities, local government and Department of Health (or equivalent) for hygiene inspection and certification,

Future plans: objectives of the business and expectations for the next 3-5 years.

2.4 Legal Aspects

2.4.1 Registration of the Enterprise

Before starting registration procedures, aspiring entrepreneurs should seek professional advice from small enterprise advisors, accountants or solicitors on what is the best type of enterprise to establish, as this varies according to the legal framework and economic structures prevailing at a given time in a particular country.

The form that a business takes is influenced by both the wishes or needs of the owner and also by the types of product that are to be made. Very frequently, new businesses are established by a family member or by a farmer who wishes to process the crops to add value. It is usually simpler and cheaper for these people to register either as personal business with unlimited liability or as a limited company with a single owner/director. However, this may not be appropriate if additional partners are required to contribute capital or specific skills.

Other types of business that can therefore be considered include a limited liability company with several directors or an un-incorporated association that has no limited liability (see Glossary for the meanings of these terms). If the proposed enterprise has a larger number of interested investors, for example a farmers' association, or if the aims also include social benefits (see Part I, Introduction), the form of the business could be a co-operative association, a not-for-profit organisation or a registered charity. However, it should be noted that charity law in many countries prohibits trading.

Once the form of the business has been decided, there are a number of registration procedures that need to be taken before it can begin trading. Again, these vary in both the number of steps and the degrees of complexity and bureauocracy in different countries and it is not possible to be comprehensive in a book of this nature. Professional advice is needed to guide the entrepreneur through each stage, from a solicitor and an accountant who are experienced in the national law in the particular country.

However, as a general guide it is necessary to register a processing business with some or all of the following authorities:

- notify the taxation authorities (e.g. the Sales Tax Commissioner or VAT Office) and complete *Notification of Business Intention* forms or their equivalent

- notify another branch of the taxation authorities to get an *Approval Certificate* to indicate that no unpaid income tax is outstanding

- apply to the Local Government Office (e.g. a Town Council or District Council) for a *Business Licence*

- apply to the Ministry of Health or Bureau of Standards, requesting that a Food Inspector visits the premises to examine the facilities

- when the inspection is completed and a satisfactory report is made, apply for registration as a Food Premises and the issuing of a *Food Producer's Licence* or *Certificate* or equivalent

- apply to the Bureau of Standards or equivalent, sending a sample of product for chemical analysis. This will be tested to determine whether it conforms to national legislation on food composition (see below). If it conforms, a *Product Approval Certificate* or similar will be issued and a Standards Authority symbol may be placed on the product label

- apply to the Ministry of Finance, Department of Customs and Local Government Tax Authority or VAT Office if there are opportunities for remission of taxes on imported ingredients, packaging materials or equipment. This is also necessary to reclaim VAT.

- register the business at a bank and open an account for trading.

If export is being considered it is also necessary to apply to:

- the Ministry of Trade and Industry or equivalent for a *Business Licence*
- the National or Central Bank for an *Export Licence*
- the Customs Department for clearance to export.

The majority of these stages require a fee to be paid and there is clearly a considerable expense in both time and travel costs to complete the registration procedure, particularly when the business is located away from the capital city. For inexperienced entrepreneurs, this is a daunting process and many either do not complete it or attempt to bypass it. However, incomplete or fraudulent registration leaves the entrepreneur open to later complications and possible prosecution.

More positively, there are schemes in many countries that are operated by National or Local Governments to encourage small enterprise development. These can vary from subsidised electricity, tax-holidays (tax-free periods of up to ten years), free land for enterprise development, support from a Small Business Advisory Service, or special bonuses if the business aims to employ women, rural people or social and ethnic minorities. The business should therefore be registered with the Ministry of Trade and Industry or other authorities that operate the schemes, in order to obtain these benefits.

Similarly, a number of international development organisations (Appendix IV) operate schemes to encourage small enterprise development and they should be contacted to find out whether credit, training or business support are available. Their local offices are often located in the Capital city, but some may have sub-offices in District Centres. Where donor countries operate bilateral development programmes, information about the available assistance can be obtained from the Aid Section of their Embassies or High Commissions. Finally the new entrepreneur should contact the local Manufacturers' Association or Chamber of Commerce to find out what assistance they can offer to get a new business established.

2.4.2 Food Related Laws

In most countries there are general laws that govern the sale of all goods, including foods, which state that any product should be suitable for its intended purpose. There are also food laws relating to the effects of food on health that say: *It is an offence for anyone to add anything to food, to process food or to sell food so that it is injurious to health with the intention that it is sold for human consumption.*

Most countries have laws to protect customers from adulteration of foods or other forms of cheating. Typically these say that *it is an offence to sell food that is not of the nature, substance nor quality demanded by the purchaser.* It is also an offence to falsely describe a food on the label or in advertising, with the intention of misleading the customer.

Within these general laws, there are also specific laws that deal with the composition of certain foods, their safety, the amount that is contained in a package, the hygiene of operators and sanitation of premises where foods are made. These laws are detailed and specific to each country, although they are often based on, or derived from, laws that have been developed in Europe or the USA during the last century. In this section it is therefore only possible to describe the broad outline of the legislation. Entrepreneurs should seek advice on specific national interpretations from professional staff in the local Bureau of Standards or from food technologists who work in universities or food research institutes. Many of these institutions have field workers or outreach programmes that are intended to provide such advice.

Figure 35. - Some headlines showing the effect of not considering food laws

Food composition

Laws relating to the composition of processed foods are complex and specific to particular types of food, particularly those such as pies or prepared foods that present opportunities for adulteration. The intention of the laws is to produce a standard for a particular food and so ensure that all foods sold with that name have that standard composition. This approach is changing in some countries and is being relaxed because of the difficulties of enforcement. Instead the authorities are relying on stricter labelling requirements to inform consumers of the food composition (see labelling below).

If export to Europe, ASEAN or USA is being considered, it is essential to obtain a detailed specification of the required product composition and quality from the local Export Development Authority, from importing companies or their agents, from Fair Trading Organisations or from the Trade Section in Embassies of the countries concerned.

It is not possible in a book of this type to detail the limits that are set for individual foods as these vary from country to country. The section below therefore describes the types of compositional standards that may be in force and the producer should find precise details of each from local authorities or Bureaux of Standards.

In relation to fruit and vegetable products, the following often have compositional standards:

Fruit juices and nectars (Sections 2.2.7 and 2.2.8):

Juice should be only pure juice with nothing added except vitamin C, specified acids used to adjust the pH and maximum levels of residual sulphur dioxide if this has been used as a preservative. Typically nectars should contain a minimum % juice, between 25% and 40% juice depending on the type of fruit and a maximum of 20% sugar or honey. There are also minimum limits for the acid content of nectars.

Soft drinks (Section 2.2.8):

Squashes, crushes and cordials are each defined in law and have minimum fruit contents specified for different types of fruit. These are typically between 1.5% and 5% minimum fruit content for drinks that are not diluted and 7% to 25% minimum fruit content for those drinks that require dilution. Dilution must be four parts water to one part drink. They each have maximum permitted levels of sugar or artificial sweeteners and can contain specified food acids.

Jams and similar products (Section 2.2.9):

Jams should contain a minimum amount of fruit pulp, which varies with the type of fruit being used, but for many is around 200g pulp per kg product. Similarly the amount of fruit juice in jelly and marmalade is specified. Normally jams should have minimum of 60% soluble solids (in practice 68%-70% is used to achieve adequate preservation, especially in tropical climates) and there are limits on residual sulphur dioxide in all products. There are detailed regulations covering definitions of the names *jams, jellies, marmalades, conserves, preserves, extra jam or jelly* and *reduced sugar jam, jelly or marmalade*.

Tomato ketchup (Section 2.2.6):

This should have a minimum of 6% tomato solids and not contain seeds. There is a maximum limit on contamination with copper and no other fruits or vegetables can be used except onions, garlic or spices for flavouring.

Additives and contaminants:

There are lists of permitted food colours, emulsifies, stabilisers, preservatives and other additives that can be added to foods. Any chemical that is not on these lists cannot be used. There are also maximum levels set for each additive in specific foods and lists of foods that are able to contain specified preservatives. Contaminants, including poisonous metals such as arsenic and lead, have maximum permitted levels in specified foods.

Food labelling

When prosecutions of food companies are analysed, a large percentage often relate to 'technical' breaches of the law because a label is incorrectly designed. It is therefore in the processors' interest to involve the local Bureau of Standards at an early stage of label design to avoid problems with prosecution and expensive re-design after labels have been printed. There are general labelling requirements that describe the information that must be included on a label (see also Section 2.8.3), but in many countries there are also very detailed laws concerning some or all of the following aspects:

- specify names that must be given to different types of ingredients
- ingredients that are exempt from the law
- the use of words such as *Best before* and *Sell by*
- the declaration of alcohol content on spirit drinks
- locations of the name of the food, the sell-by date and the net weight (they must all be in the same field of vision when a customer looks at the label)
- the visibility of information and the ability of customers to understand it (including the relative print sizes of different information)
- claims and misleading descriptions, especially about health-giving or tonic properties, nutritional advantages, diabetic or other medicinal claims
- specifications of the way in which certain words such as *flavour*, *fresh*, *vitamin* etc. can be used.

This is also a complex area, which is not possible to describe in detail in this book and professional advice should be sought from graphic designers who are experienced in label design, or from a Bureau of Standards.

Hygiene and sanitation

Laws relating to food production premises and the staff who handle foods are among the most widely enforced in most developing countries. There are numerous examples of prosecutions by Food Inspectors from Ministries of Health or other enforcement authorities and in some cases, enforced closure of the business for failure to comply with these laws. Guidelines on the design and construction of premises (Section 2.5.3) and hygiene of operators (Appendix I) should therefore be consulted before submitting a new processing facility for inspection and certification (see above).

These guidelines should be rigorously enforced in routine production to ensure that safe, high quality products are produced (see also Section 2.7.2). In summary the laws are concerned with the following aspects of health, hygiene and sanitation:

- processing that is carried out in unsanitary conditions
- or where food is exposed to the risk of contamination
- equipment (which must be able to be cleaned and kept clean)
- persons handling food and their responsibilities t
  to protect it from contamination
- building design and construction including water supplies,
- drainage, toilet facilities, wash-hand basins, provision of first aid facilities, places to store clothing, facilities for washing food and equipment, lighting, ventilation, protection against infestation by rats and insects and removal of wastes.

Weights and measures

The aim of this type of legislation is to protect customers from being cheated by unscrupulous manufacturers, for example from being sold underweight packs of food. The laws are to ensure that the amount of food that is declared on the label as the net weight (the weight of product in a pack) is the same as the weight of food that is actually in the pack. However, it is recognised that not every pack can be filled with <u>exactly</u> the specified weight because both machine-filling and hand-filling of containers is subject to some variability. The laws are therefore designed to allow for this variability but to prevent fraud.

There are two types of weights and measures legislation in force in different countries: the older method, which is still used in most developing countries, is known as the *Minimum Weight System*. This is intended to ensure that every pack of food contains at least the net weight that is written on the label. If any pack is found below this weight the producer is liable for prosecution. This system works well to protect customers, but is more expensive for producers because they have to routinely fill packs to just above the declared weight to avoid prosecution and they therefore give a small amount of product away in every pack.

A second type of legislation was introduced in Europe to take account of the automated filling and packaging that is used by most producers there. This is known as the *Average Weight System* and uses a statistical probability of a defined proportion of packages being above the declared weight as a basis for enforcement. As most small scale producers in developing countries do not use automatic fillers and programmable check-weighers, this system is difficult to operate and un-necessarily complex. If however, a producer is considering export to an industrialised country, advice and information on this legislation should be obtained from a local Export Promotion Board or equivalent institution so that an *'e' mark* can be obtained to indicate that the process conforms to this system.

In some countries, there are specified weights that must be used when selling dried fruits and vegetables and jams or marmalades (but not other processed fruit and vegetable products).

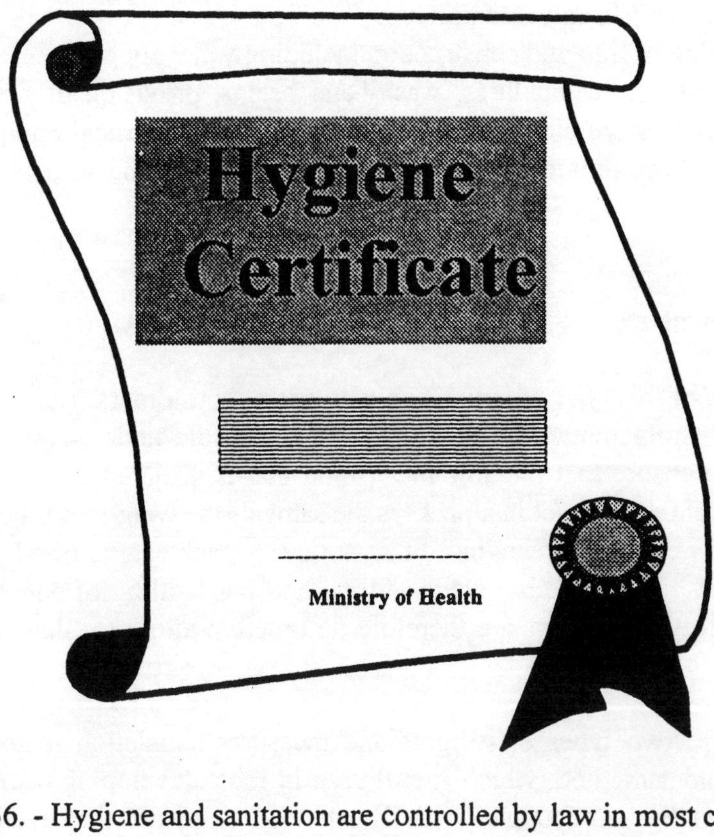

Figure 36. - Hygiene and sanitation are controlled by law in most countries

2.5 Establishing Production Facilities

2.5.1 Introduction

All fruit and vegetable processing operations require an hygienically designed and easily cleaned building to prevent products from becoming contaminated during processing. The two main sources of contamination are 1) insects and animals and 2) micro-organisms. Insects and animals are attracted to food buildings if foods or wastes are left lying round after production has finished. Micro-organisms can grow in food residues that are left on equipment, tables or floors which have not been properly cleaned. Micro-organisms require water to grow (Section I, Introduction) and wet processing therefore has an inherently greater risk of contamination than dry processing does. However, some types of micro-organism can form inert spores that are able to survive under dry conditions and then grow when they come into contact with water or foods and strict hygiene should also be enforced in drying operations. In dry processing there is an additional risk of contamination by dust, which can spoil foods itself and also harbour micro-organisms. The following aspects of setting up a processing facility should therefore be addressed by entrepreneurs, whether they are constructing a new facility or converting an existing building.

2.5.2 The Site

The location of a food building is very important and the following aspects need to be considered when choosing a site:

- location in relation to raw material supplies and likely markets
- ease of access for staff (public transport, distance down an access road)
- quality of road access (all year, dry season only, potholes that may cause damage to products, especially when glass containers are used)
- nearby swamp land that would be a source of smells and insects
- any potential contamination of water supplies upstream of the processing site
- available land for waste disposal away from the building
- electricity supplies
- cleared land to reduce problems caused by insects and birds (preferably planted with short grass, which acts as a dust trap for airborne dust).

2.5.3 Design and Construction of the Building

In general, a building should have enough space for all production processes to take place without congestion and for storage of raw materials, packaging materials and finished products. However, the investment should be appropriate to the size and expected profitability of the enterprise to reduce start-up capital, the size of any loans taken out and depreciation and maintenance charges.

Roofs and ceilings

In tropical climates, overhanging roofs keep direct sunlight off the walls and out of the building. This is particularly important when processing involves heating, to make working conditions more comfortable. Fibre-cement tiles offer greater insulation against heat from the sun than galvanised iron sheets do. High level vents in roofs both allow heat and steam to escape and encourage a flow of fresh air through the processing room. The vents must be screened with mesh to prevent insects, rodents and birds from entering the room. If heat is a serious problem, the entrepreneur could consider fitting electric fans or extractors, although this clearly increases capital and operating costs.

Rafters or roof beams within the processing and storage rooms are unacceptable. They allow dust to accumulate, which can fall off in lumps to cause gross contamination of products. Similarly, insects can fall from them into products. They also allow paths for rodents and birds, with consequent risks of contamination from hairs, feathers or excreta. It is therefore essential to have a paneled ceiling fitted to any processing or store-oom, with careful attention when fitting them to ensure that there are no holes in the paneling. Care should also be taken to prevent birds, rodents and flying insects gaining access to the processing room through gaps in the roof structure or where the roof joins the walls.

Walls

As a minimum requirement, all internal walls should be rendered or plastered with a good quality plaster to prevent dust forming in the processing room. An experienced plasterer should be used to ensure that no cracks or ledges remain in the surface finish, which could accumulate dirt and insects. The lower area of walls, to at least 1.08 metres (four feet) above the floor, is most likely to get dirty from washing equipment, from product splashing etc. and special attention should be paid to ensure that this area is easily cleaned. Higher areas of walls should be painted with a good quality emulsion. The lower parts of walls should be either painted with a waterproof gloss paint, preferably white, to allow them to be thoroughly cleaned, or ideally they should be tiled with glazed tiles. If tiling a process room is too expensive, it is possible to select particular areas such as behind sinks or machinery and only tile these parts. In some countries there is a legal requirement for specified internal finishes and this should be checked with the Ministry of Health or other appropriate authority (see also Section 2.4.2)

Windows and doors

Window sills should be made to slope for two reasons: to prevent dust from accumulating and to prevent operators from leaving cloths or other items lying there, which in turn can attract insects. Windows allow staff to work in natural daylight, which is preferable to and cheaper than electric lighting. However, in tropical climates there is a natural inclination for workers to open windows to allow greater circulation of fresh air. This provides easy access for flying insects, which can readily contaminate the product. Windows should therefore be fitted with mosquito mesh to allow them to be left open.

Normally doors should be kept closed, but if they are used regularly there is again a tendency for them to be left open with similar consequences of animals and insects entering the plant. In this case, thin metal chains or strips of material that are hung vertically from the door lintel may deter insects and some animals, while allowing easy access for staff. Alternatively mesh door screens can be used. Doors should be fitted accurately so that there are no gaps beneath them and all storeroom doors should be kept closed to prevent insects and rodents from destroying stock or ingredients.

Figure 37 - A well-designed processing room

Floors

It is essential to ensure that the floors of processing rooms and storerooms are constructed of good quality concrete, smooth finished and without cracks. In some developing countries, it is possible to buy proprietary floor paints or vinyl based coatings, but these are usually very expensive. Generally, it is not adequate to use the red wax floor polishes that are commonly found in households, as these wear away easily and could contaminate either products or packages. Over time, spillages of acidic fruit products react with concrete and cause it to erode. Attention should therefore be paid to cleaning up spillages as they occur and to regularly monitor the condition of the floor.

The corners where the floor and the walls join are places for dirt to collect. During construction of the floor, it should therefore be curved up to meet the wall. It is possible to place fillets of concrete (or 'coving') in the corners of an existing floor to fill up the right angle, but care is needed to ensure that new gaps are not created which would harbour dirt and insects.

The floor should slope at an angle of approximately 1 in 8 to a central drainage channel. At the end of a day's production, the floor can be thoroughly washed and drained. Proper drainage prevents pools of stagnant water forming, which would in turn risk contamination of equipment and foods. The drainage channel should be fitted with an easily removed steel grating so that the drain can be cleaned. Where the drain exits the building, there is a potential entry point for rodents and crawling insects unless wire mesh is fitted over the drain opening. This too should be easily removed for cleaning.

Lighting and power

General room lighting should be minimised wherever possible. Full use should be made of natural daylight, which is both free and better quality light, especially for intricate work. Where additional lighting is needed, florescent tubes are cheaper to operate than incandescent bulbs. However, if machinery is used that has fast moving exposed parts, these should be lit with incandescent bulbs and not tubes. This is because even though the parts should have guards fitted, a rotating machine can appear to stand still if its speed matches the number of cycles of the mains electricity that powers fluorescent tubes - with obvious dangers to operators.

All electric power points should be placed at a sufficiently high level above the floor that there is no risk of water entering them during washing the floor or equipment. Ideally, waterproof sockets should be used. It is important to use each power point for one application and not use multiple sockets which risk overloading a circuit and causing a fire. If there are insufficient power points for the needs of a process, additional points should be installed, even though this is more expensive. All plugs should be fitted with fuses that are appropriate for the power rating of the equipment and ideally the mains supply should have an earth leakage trip switch. If three-phase power is needed for larger machines or for heavy loads from electric heating, it is important that the wiring is installed by a qualified electrician to balance the supply across the three phases.

Water supply and sanitation

Water is essential in nearly all fruit and vegetable processing, both as a component of products and for cleaning. An adequate supply of potable water should therefore be available from taps around the processing area. In many countries, the mains supply is unreliable or periodically contaminated and it is therefore necessary for the entrepreneur to make arrangements to secure a regular supply of good quality water each day. This can be done by installing two high level, covered storage tanks either in the roof-space or on pillars outside the building. They can be filled alternately when mains water is available and while one tank is being used, any sediment in water in the other tank is settling out. As sedimentation takes several hours, the capacity of each tank should be sufficient for one day's production. The tanks should have a sloping base and be fitted with drain valves above the slope and at the lowest point. In use, water is taken from the upper valve and when the tank is almost empty, the lower valve is opened to flush out any sediment that has accumulated.

Water that is included in a product should be carefully treated to remove all traces of sediment and if necessary, it should be sterilised. This is particularly important if the product is not heated after water has been mixed in as an ingredient.

There are four ways of treating water at a small scale: by filtration; by heating; by ultra-violet light and by chemical sterilants, such as hypochlorite (also named 'chlorine solution' or 'bleach'). Other water treatment methods are generally too expensive at a small scale of operation.

Filtration through domestic water filters is slow, but having made the capital expenditure, it is relatively cheap. Larger industrial filters are available in some countries. Heating water to boiling and holding it at that temperature for 10-15 minutes is simple and has low capital costs, but it is expensive because of fuel costs and it is time consuming to do routinely. Heating sterilises the water but does not remove sediment and boiled water may therefore require filtering or standing to remove sediment.

Ultra-violet light destroys micro-organisms in water and commercial water treatment units that use this principle (Figure 38) are coming down in price to the point that they can be suitable for those small scale processors that use a lot of water. Again, this method does not remove sediment from the water.

Finally, chemical sterilisation using hypochlorite is fast, relatively cheap and effective against a wide range of micro-organisms. Cleaning water should contain about 200ppm of chorine and water that is used as an ingredient should contain about 0.5ppm to avoid giving a chlorine flavour to the product. A chlorine concentration of 200ppm can be made by adding 1 litre of bleach to 250 litres of water and a 0.5ppm solution is obtained by adding 2.5 ml of bleach to 250 litres of water. Although chlorine kills most micro-organisms, it also has a number of disadvantages: it can corrode aluminium equipment; it can taint foods; bleach must be handled with great care as it damages the

necessary, the concentration of chlorine in water can be measured using a chemical dye that produces a colour when it reacts with chlorine. The intensity of the colour is compared to standard colours on glass discs in a 'comparator'.

Good sanitation is essential to reduce the risk of product contamination and to deter insects, rodent and birds. All wastes should be placed in bins and not piled on the floor. Processes should have a management system in place to remove wastes from the building as they are produced, rather than letting them accumulate during the day. Wastes should never be left in a processing room overnight. This aspect is described further in Section 2.7.2, and summarised in Appendix I.

Figure 38. - Ultra-violet water steriliser
(Courtesy of UV Systems Ltd.)

Good sanitation is essential to reduce the risk of product contamination and to deter insects, rodent and birds. All wastes should be placed in bins and not piled on the floor. Processes should have a management system in place to remove wastes from the building as they are produced, rather than letting them accumulate during the day. Wastes should never be left in a processing room overnight. This aspect is described further in Section 2.7.2, and summarised in Appendix.

Layout of equipment and facilities

The different areas required for fruit and vegetable processing are shown in Figure 39 for a drying unit and in Figure 40 for other types of production. The layouts of these processing rooms show how raw materials move through a process and through the room without paths crossing. Different stages in a process should be physically separated wherever possible. This helps prevent contamination of finished products by incoming, often dirty, raw materials and clearly identifies areas of the room where special attention to hygiene is necessary. This is particularly important to prevent contamination arising from activities such as bottle washing in which inevitable breakages produce glass splinters that could contaminate a product. This separation also reduces the likelihood of accidents or of operators bumping into each other.

Perishable raw materials should be stored separately from non-perishable ingredients and packaging materials. A separate office allows records to be filed and kept clean and provides a quieter working environment for book-keeping. Toilets should either be housed in a separate building or two doors should exist between them and a processing area. All workers should have access to hand-washing facilities with soap and clean towels. Laboratory facilities are generally not needed in fruit and vegetable processing, although a separate table for conducting quality assurance checks or check-weighing packages of finished product (Section 2.7.2) could be located in the office or in a separate area of the processing room.

2.5.4 Equipment

When selecting equipment, it should be the correct size for the intended scale of production (obtained from the Feasibility Study, Section 2.3.3). Managers should devise regular maintenance and cleaning schedules and ensure that they are followed. Further details on these topics are given in Section 2.7.1, Production Management and Section 2.7.2, Quality Assurance.

All types of fruit and vegetable processing require basic equipment to handle, weigh and prepare raw materials, such as buckets, tables, knives, and scales. Ideally, two sets of scales are used, one with an accuracy of +/- 0.1g to accurately weight small amounts and a second having an accuracy of +/- 50g for larger amounts of raw materials. However, scales are expensive to buy in most countries although the cost of small, electronic domestic scales is falling (Figure 41). A cheaper alternative to buying scales is to calibrate scoops or other measures, so that they contain the correct quantity of material when filled level with the top. In operation, scoops are faster than weighing, but the level of accuracy may be lower and careful training of operators is needed to ensure that the weights are consistent.

Because of the acidic nature of fruits, the parts of equipment that are in contact with foods should be made from either food grade plastic, aluminium or stainless steel. Other metals, such as mild steel, brass and copper should not be used because they react with the fruit and cause off-flavours or colour changes in the product. In general, because of its high cost, stainless steel is only used for cutting blades, boiling pans etc..

Wooden tables are cheaper in most countries than metal ones, but they are more difficult to keep clean. Ideally, wood should be covered in a sheet of thick plastic, aluminium or a 'melamine' type surface for easier cleaning. Details of methods for hygienic handling and storage of equipment are given in publications in the Bibliography and are summarised in Appendix I.

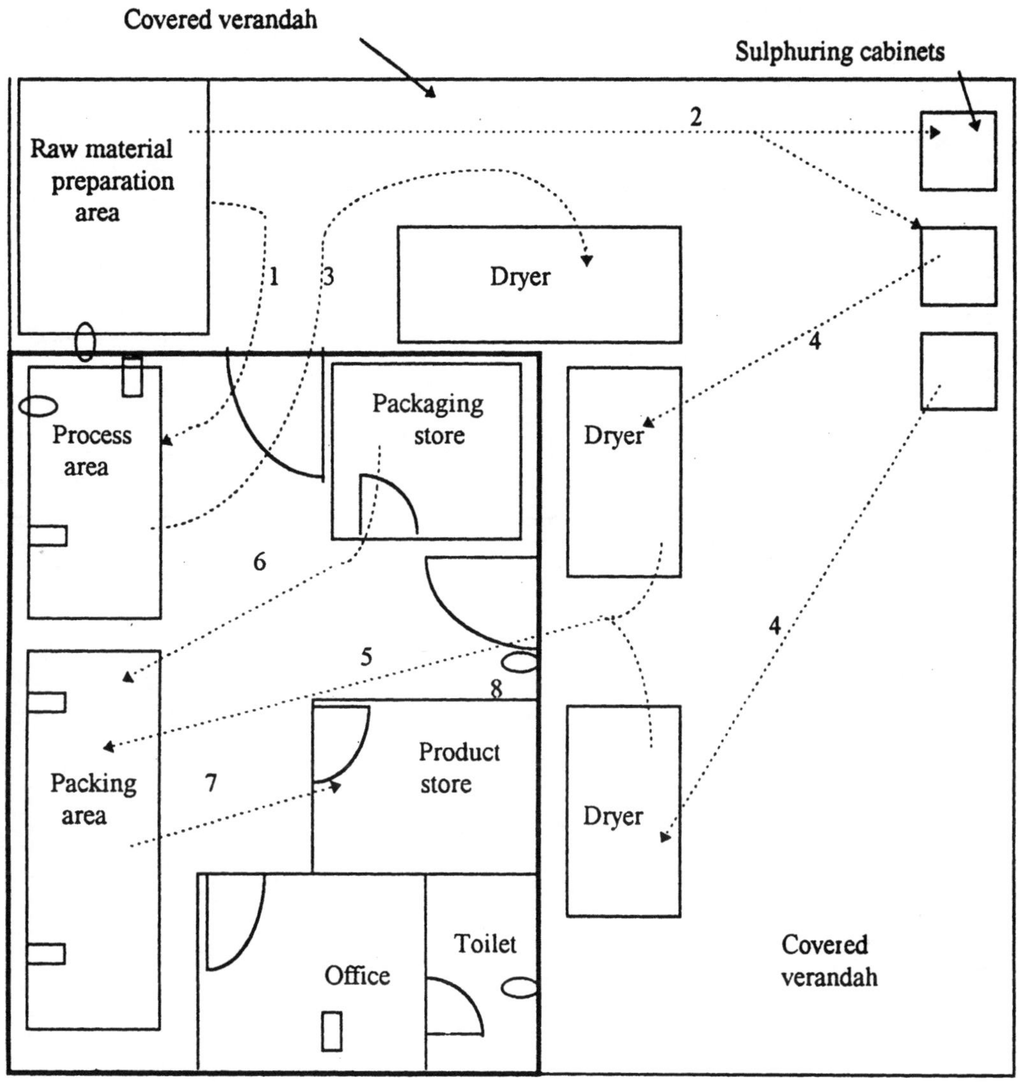

Figure 39. - Layout of a building for drying of fruit and vegetables

Key:

1 Prepared raw materials for processing
2 Prepared raw materials for sulphuring
3 Blanched/crystallised food to dryers
4 Sulphured fruits to dryers
5 Dried foods for packing
6 Packaging from store
7 Packaged product to store for distribution
8 Hand washing

□ = power point

0 = water tap

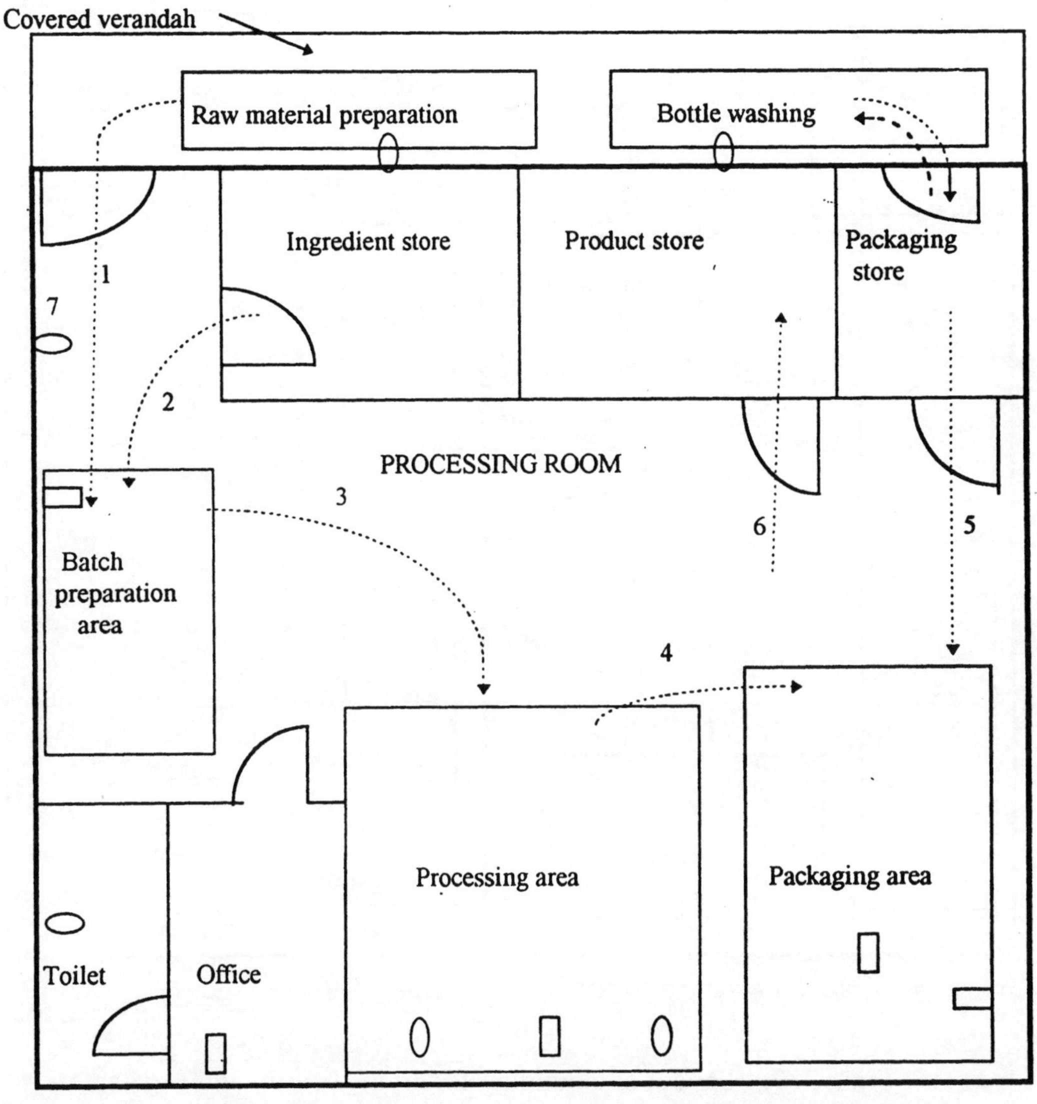

Figure 40. - Layout of a building for wet processing of fruit and vegetables

Key:
1 Prepared raw materials in
2 Ingredients from store
3 Prepared batch to processing area
4 Processed food to packaging area
5 Packaging from store
6 Packaged product to store for distribution
7 Hand washing

▭ = power point
0 = water tap

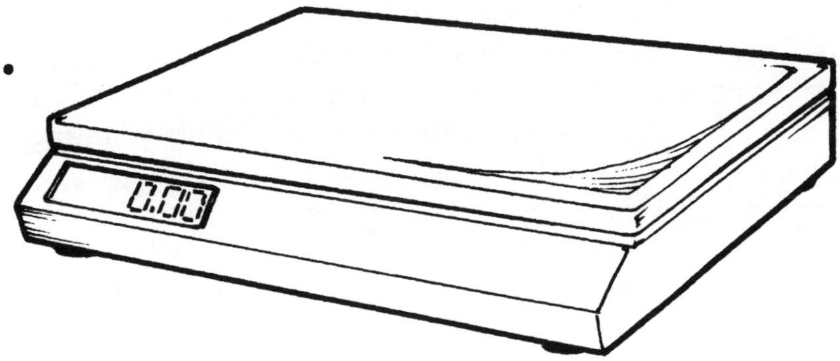

Figure 41. - Small electronic weighing scales

Dried products

The process flow chart for dried foods (Figure 10) indicates a number of processing options depending on the type of product that is made. For example vegetables are frequently blanched by placing them in a wire basket and dipping them into a pan of boiling water (Figure 42). Alternatively they can be placed on wire mesh in a steam chamber. Fruits are often sulphured or sulphited to protect their colour. Sulphite dips are contained in a tank made from food grade plastic and sulphuring cabinets, comprising a wooden box, fitted with mesh trays, can easily be made from locally available materials.

Figure 42. - Blanching vegetables before drying

At larger scales of operation, there are a range of machines that can be used for preparation of fruits and vegetables. These include cleaners, de-stoners, peelers, cutters and slicing or dicing equipment (for example Figure 9). Details of equipment and suppliers are given in publications in the Bibliography. In the following Section, the types of equipment that are used to make dried fruits and vegetables, boiled and pasteurised products and wines, vinegars and spirits are described.

Fruits for crystallising are soaked in syrup using food grade plastic tanks and aluminium pans for boiling the syrup. A series of tanks are used to gradually increase the concentration of syrup over 3-4 days (Figure 11), which also allows greater utilisation of sugar compared to single stage soaking. In most countries, sugar contains dust and other contaminants and syrups should therefore be filtered through muslin cloth before use.

A number of different types of drier can be used, depending on the level of investment that can be justified from the feasibility study. A comparative summary of cabinet dryers in relation to other types of dryer is shown in Table 23. Solar dryers have some advantages over sun drying when correctly designed. They give faster drying rates by heating the air to 10-30°C above ambient, which causes the air to move faster through the dryer, reduces its humidity and deters insects.

Type of Dryer	Cost ($)	Capacity (kg wet food per day)	Investment ($ per kg of dry food)	Fuel efficiency	Labour requirement
Brace solar	50	10	50	n/a	very low
Solar cabinet	70	30	23	n/a	very low
McDowell	170	40	43	very poor	low
Wood burning	340	80	43	very poor	low
ITDG fuel fired batch	3,400	240	140	poor	high
ITDG fuel fired semi-continuous	6,800	360	190	medium	very high
Small commercial cabinet	85,000	500	1,700	good	high
Large commercial cabinet	170,000	2,500	680	good	medium
Tunnel (12 carriage)	145,000	6,000	240	good	low
Moving band	800,000	48,000	170	very good	very low

n/a - not applicable

Table 23. - A comparison of different drying technologies
(From *Try Drying IT*, by Axtell and Bush, courtesy of IT Publications)

The faster drying reduces the risk of spoilage, improves the quality of the product and gives a higher throughput, so reducing the drying area that is needed. However, care is needed when drying fruits to prevent too rapid drying which would result in case hardening (see glossary) and subsequent mould growth. Solar dryers also protect foods from dust, insects, birds and animals. They can be constructed from locally available materials at a relatively low capital cost and there are no fuel costs. They may therefore be useful

1) where fuel or electricity are expensive, erratic or unavailable,
2) where land for sun drying is in short supply or expensive,
3) where sunshine is plentiful but the air humidity is high, and
4) as a means of heating air for artificial dryers to reduce fuel costs.

This last application is likely to gain in importance as fuel prices increase or to reduce dependence on imported fuels. Solar drying is not likely to be useful where the quality of sun dried foods is acceptable to local consumers and where the additional costs of solar drying are not recovered from increased value of the food.

In situations where the control over drying conditions is insufficient using solar dryers, it is necessary to use a fuel-fired dryer. Again, there are a large number of different types and the selection depends on the required throughput, types of fuel and level of investment that are available. The main limitations of fuel-fired dryers, in addition to higher capital and operating costs, are that they are more complex to build and maintain and therefore require skilled labour for operation and maintenance. There are many designs of each type, which are described in detail in publications in the Bibliography.

One type of dryer that has found application in many developing countries is the cabinet dryer (Figure 43), which is successfully used for drying herbs, herbal teas and spices and is also suitable for fruits and vegetables. In general a drying area of $1m^2$ is needed for 2-6kg of raw materials, depending on the type of food (6kg of chopped fruits need $1m^2$ whereas a product like shredded cabbage is less dense and can only be stacked at around $2kg/m^2$). These figures allow calculation of the size of the dryer that is needed for a given weight of food to be dried per day.

Boiled, concentrated and pasteurised products

This range of products includes juices, squashes, sauces, pickles and chutneys. For liquid products, juice or pulp can be extracted from fruits or vegetables in a number of ways, depending on the hardness of the raw material. Soft fruits and vegetables, such as berries, tomatoes, grapes etc. can be processed by pressing, using a fruit press or using a juicer attachment to a food processor (Figure 44). Citrus fruits are usually reamed (Figure 45) to extract the juice without the bitter pith or skin.

Passion fruit or tomato and harder fruits, such as apple, pineapple etc. are peeled and then pulped using a liquidiser, or at large scales of operation using a pulper-finisher

(Figure 16), which separates skins and seeds. Steamers, such as those used for blanching, can also be used to 'dissolve' some types of cut soft fruits such as melon and pawpaw. When a clear juice is required it is necessary to filter it through a fine muslin cloth or stainless steel juice strainers.

Figure 43.- -A batch fuel fired dryer (Courtesy of Midway Technology)

Figure 44. - Small juice extractor, fitted to a food processor
(Courtesy of Midway Technology)

The majority of products require heating to either pasteurise or concentrate them. In all cases a stainless steel boiling pan is needed. These are expensive to buy and in many countries local fabrication is difficult because the skills and facilities for welding stainless steel are not readily available. However, there are few alternatives and a producer should regard this expenditure as a necessary investment to be able to produce a high quality product. In some cases, such as squash production, it is possible reduce the capital investment by heating syrup to boiling in a large pan made from cheaper aluminium and then mixing it with juice in a smaller stainless steel pan for a final short heating to achieve the required pasteurisation conditions.

There are two types of boiling pans available, depending on the scale of operation: at smaller scales of operation, a simple stainless steel pan can be placed directly over the heat source. At larger scales of production, an indirectly heated, or 'double jacketed' pan (Figure 18) can be used. Steam is produced by a boiler and fed into the space between the outer jacket and inner pan to give more uniform heating and therefore avoid localised burning of the product. This may be particularly important when heating viscous products such as sauces, jams, chutneys etc., which are more likely to stick to a simple pan and burn onto it. This would not only reduce the quality of the product but also significantly slow down production while the pan is cleaned between batches.

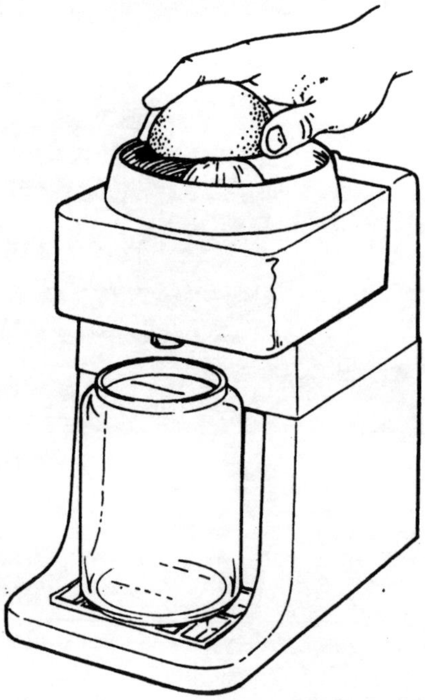

Figure 45. - A small scale electric citrus reamer

The most appropriate type of heat source depends on the cost and availability of different fuels where the production is located. In general, gas or electricity are the preferred options because there is no risk of contamination of the product by un-burned fuel or combustion gases. A comparison of the advantages and limitations of different fuels is shown in Table 24, but the final decision on which fuel to use is likely to be based on considerations of finance and availability during the feasibility study. The majority of products also require equipment for monitoring and control, particularly those in which the concentration of sugar, salt or acid is important for preservation. This is discussed in more detail in Section 2.7.2.

Fermented and distilled products

In addition to the equipment required to prepare juices for fermentation, this group of products require more specialised equipment for fermentation and distillation. Wines are fermented in either food grade plastic drums or large glass vessels, which have a narrow opening into which an air lock (Figure 24) can be fitted. An alcohol hydrometer (similar in appearance to the brine hydrometer in Figure 55) is not essential, but it is a useful aid to standardising the alcohol content of the products.

It is possible to make vinegar by simply exposing wine to the air, but yields are low and there is a high risk of spoilage. A commercially made vinegar fermenter is too expensive for most producers, but if expertise is available, a locally produced fermenter having a traditional design can be made (Section 2.2.10).

Commercially produced distillation apparatus is also expensive to buy and locally made alternatives are likely to be preferred by small scale producers. With experience and adequate control over heating, these can produce acceptable products. They should be fitted with a pressure safety device, such as a long pipe that is submerged below the level of the liquor and exits the heating vessel to a height of at least 1.5 metres. If the outlet to the still becomes blocked, this will prevent the pressure rising to the point where the still would explode.

Packaging, filling and sealing equipment

All types of plastic film, with the exception of un-coated cellulose, can be sealed using a heat sealer (Figure 46). The differences in the types of sealer are due to the width of the heated bar or wire and the level of control over temperature and time of heating. For dried and liquid foods a relatively wide seal (e.g. 3-5mm) is required and bar-type sealers are therefore preferable to wire-types. The sealer should also have a thermostat to adjust the sealing temperature, and an adjustable timer to control the time of heating. Care should be taken to ensure that there is no product dust on the inside of the package where the seal is to be made as this will prevent proper sealing.

Criteria	Electricity	Gas	Liquid fuels	Solid fuels
Energy per unit weight or volume[a]	not applicable	low	high	moderate to high
Cost per unit of energy[b]	moderate to high	high	moderate to high	low
Heating equipment cost	low	low	high	high
Efficiency of heating	high	moderate to high	moderate to low	low
Flexibity of use	high	high	low	low
Fire or explosion hazard	low	high	low	low
Risk of contaminating food	low	low	high	high
Labour and handling cost	low	low	low	high

[a] Heating values (in kJ/kg x 10^3) for gas = 1.17-4.78, for oil = 8.6-9.3, for coal = 5.26-6.7, for wood = 3.8-5.26.

[b] depending on presence of national hydro-electric schemes, coal mines or afforestation projects

Table 24. - A comparison of different sources of heat for processing
(From: *Food Processing Technology*, by Fellows)

Solid products, such as pickles and chutneys are usually filled by hand using scoops or ladles into jars, plastic pots or bags. This is a time-consuming operation, which may require a large staff input (e.g. Figure 33). However, in most small scale operations, this is the only realistic option because mechanical fillers for these types of product are prohibitively expensive and usually operate at too high a throughput.

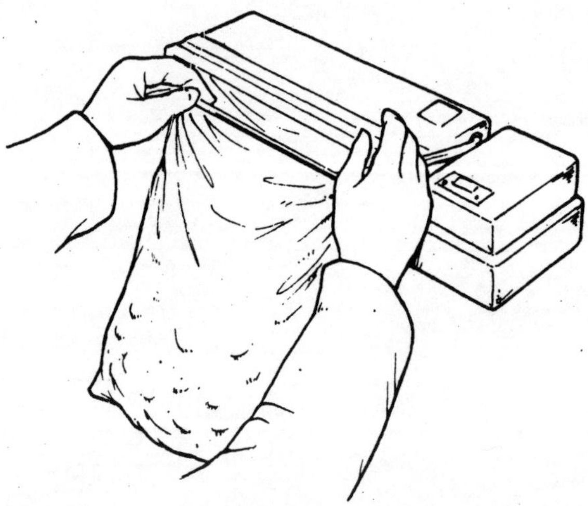

Figure 46. - A heat sealer for sealing plastic films

Although liquid products can also be filled by hand using jugs or ladles, in contrast to solid products there are a number of small liquid fillers available, which are affordable by many small scale producers. Examples of these include gravity fillers, made by fitting gate valves to stainless steel or food grade plastic tanks (note: domestic taps should not be used because they are too difficult to clean properly). Other designs include volumetric fillers and dispensers, in which a measured amount of liquid is filled into each container by the action of a piston (Figure 47). Small machines are available to seal jars, bottles, cans, plastic pots and films. Details of the various designs and suppliers are given in publications in the Bibliography.

2.5.5 Packaging Materials

There is a very wide range of packaging materials that can be used for foods and these cannot be described in detail in a book of this size. Publications in the Bibliography describe some types of packaging in more detail, but entrepreneurs should contact packaging manufacturers or their agents for a complete list of the available types. The following is a brief description of some of the more important points concerning the most widely available types of packaging materials in most developing countries. Additional information that relates to specific products is given in Section s 2.2.3 and 2.5.4. Marketing aspects are described in Section 2.8.3.

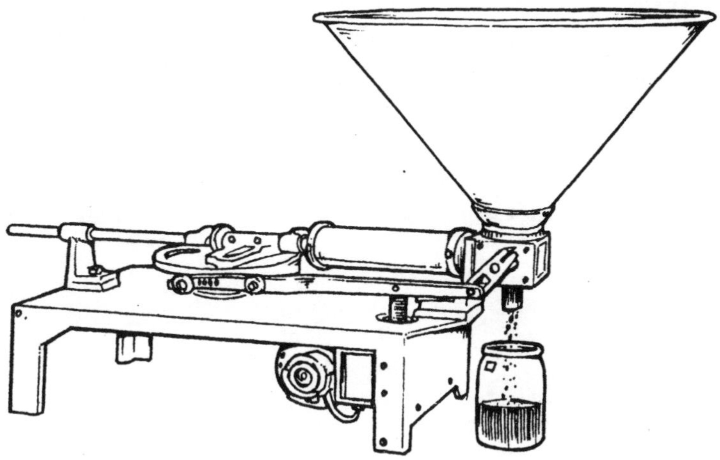

Figure 47. - A volumetric piston filler for liquid products

Jars and bottles are available in countries that have a glass-works, or have access to an overland supply from a neighbouring country. Because of their heavy weight, high bulk and fragility, glass containers are expensive to transport long distances and are frequently not available to producers in developing countries. Where they are available, they are usually re-used and great care is needed to ensure that they are properly cleaned. New and re-used containers should be sealed with new caps, lids or corks in order to obtain an adequate seal. The most common jar lids are now TOTO type (twist on, twist off), although the 'Omnia' type is still found in many countries. Bottles may be sealed using ROPP (roll on pilfer-proof) caps or corks. Cans are not widely used in small scale processing for the reasons described in Section 2.2.2.

Plastic pots and bottles are suitable for some types of foods and they are becoming increasingly common as a result of their lower production and distribution costs. Pots can be either heat sealed with a foil lid or with a snap-on plastic lid. The most common types of plastic film in developing countries are polythene and polypropylene, although increasingly there are agents who can supply more sophisticated (and expensive) imported laminates (Table 25).

Small laminated plastic/foil/cardboard cartons for UHT juices are appearing in many countries, but these are usually imported under licence to large scale juice manufacturers and are not available to small scale processors. Additionally, the UHT technology is not suitable for small scale production. Other cardboard and paper packaging is more widely available and can usually be printed by local print companies.

Other, more traditional types of packaging such as leaves, jute, hessian, wood and pottery are not usually able to convey an image of 'modern' or hygienic products and except for some niche export or tourist markets, these are not widely used (see also Section 2.8.3).

The properties of some of the more commonly used packaging films are shown in Table 25.

Type of film	Thickness (µm)	Yield ($m^2 kg^{-1}$)	Moisture trans-misson[a]	Oxygen trans-mission[b]	Strength[c]	Light transmission (%)	Sealing temperature (°C)
Cellulose							
- uncoated	21-40	18-30	800-1500	8-10	33	TP	N/A
metallized & PVdC coated	21-42	17-31	4-5	2-3	28-60	0	90-130
Polyethylene							
- low density	25-200	5-43	14-19	8000	7-16	-	121-170
- stretch wrap	17-38	-	-	-	-	-	N/A
- high density	350-1000	11-43	4-6	500-2000	24-61	-	135-170
Polypropylene							
- biaxially oriented	20-40	27-55	3-7	2000	118-260	TP	117-124
- PVdC coated	18-34	30-53	4-8	-	-	TP	120-145
- metallized	20-30	36-55	1	-	215	0.5	120-145
Polyester							
- plain	12-23	31-59	20-40	53-110	-	87	100-200
- metallized	-	-	0.8-2	0.5-1.5	-	0	100-200
Polyvinylidene chloride	10-50	17-35	1-4	2	120-130`	90	100-160

[a] Moisture vapour transmission rate (ml m^{-2} per 24 hours) measured at 38°C and 90% relative humidity

[b] Oxygen transmission rate (ml m^{-2} per 24 hours) measured at 25°C and 45% relative humidity

[c] Tensile strength (MN m^{-2}) in machine direction.

PVdC = Polyvinylidene Chloride

TP - transparent,

N/A - not applicable

Table 25. - Properties of selected packaging materials
(From: *Food Processing Technology*, by Fellows)

2.6 Contracts with Suppliers and Retailers

2.6.1 Securing Raw Materials

Many small scale processors buy fruits and vegetables daily from their nearest public market. Although this is simple and straightforward, it creates a number of problems for a business: for example, the processors have little control over the price charged by traders each day and because of the large seasonal price fluctuations that characterise these raw materials, this makes financial planning and control over cashflow more difficult (Section 2.3.4). The processor is also unable to schedule the raw materials in the quantities required and it is common for production to fail to meet a target because there are simply not enough fruits and vegetables for sale on a particular day. Additionally, the processor has no control over the way fruits and vegetables are handled during harvest and transport to the markets and therefore no influence over the quality of the raw materials that are available (see also Section 2.7.2).

To address these problems, a processor can arrange contracts with either traders or farmers, in an attempt to have greater control over the amount of raw materials available for processing each day and their quality and price. This is not a common arrangement at present in most developing countries, possibly because commercial food processing is a relatively recent activity and there is no history of collaboration and formal contracts. However, where this has been done, there are benefits to both processor and suppliers, provided that the arrangements are made honourably and there is mutual trust. The benefits to farmers are a guaranteed price for their crop, based on a sliding scale of quality and a guaranteed market when it is harvested.

However, the traders who tour an area to buy crops provide a number of benefits to farmers that processors should not ignore when arranging contracts: for example the traders frequently buy the whole crop, regardless of quality and either sort it themselves for different markets or sell it on to wholesalers who do the sorting. From the farmers' perspective, they receive payment at the farm, without having to worry about marketing their crop or disposal of substandard items. Although farmers have a 'guaranteed' market by selling to traders, they have virtually no control over the prices offered and can be exploited, particularly at the peak of a growing season when there is an over-supply of a particular crop.

Traders also provide a number of other services that farmers may find difficult to obtain elsewhere: traders may be the only realistic source of farming tools and other inputs such as seeds; they are also a source of immediate informal credit, which farmers may require to buy inputs or for other needs such as funerals and weddings. Although the interest payments on such loans may be much higher than those charged on commercial loans, farmers often have no access to banks or other lenders and in practice have no choice. In many countries, large numbers of farmers are permanently indebted to traders for their lifetimes and are only released from the debt by sale of land.

When processors begin to negotiate contracts with farmers, they should therefore be aware that farmers may be unwilling to break the existing arrangements with traders, either because of genuine fears that they will lose the services provided or because they are indebted to traders and have no ability to make other arrangements. The local power of traders should not be under-estimated and may range from a refusal to offer further loans to farmers, a threat not to buy the crop again if sales are made directly to processors, a demand that farmers repay loans immediately and in extreme cases, physical violence.

Despite the problems described above, there are possibilities for processors to agree contracts to supply fruits and vegetables of a specified variety and quality with individual farmers or with groups of farmers who may be working cooperatively.

Typically a specification would include the variety to be grown, the degree of maturity at harvest, freedom from infection etc. The price paid for the crop is agreed in advance and may be set between the mid-season lowest point and the pre- and post-season high points. Alternatively a sliding scale of prices is agreed, based on one or more easily measurable characteristics such as minimum size or agreed colour range, with an independent person being present to confirm the agreement in case of later disputes. The agreement may also specify the minimum or maximum amount that will be bought. In a formal contract, these agreement are written down and signed by both parties, although such formal contracts are rare in most developing countries.

Figure 48. - Traders are important buyers of fruit and vegetables

Processors should also consider the other forms of assistance that could be offered to farmers. For example, in some other larger scale processing such as tea and coffee production, processors offer training and an extension service to address problems with the crop as they arise throughout the growing season. Although this may be beyond the resources of small scale processors, more limited types of assistance may include purchasing tools, fertilizer or other requirements in bulk with the savings being passed on to farmers. Alternatively, part-payment for the crop can be made in advance so that farmers can buy inputs without the need for credit and the consequent indebtedness.

The advantages to the processor are greater control over the quality of raw materials and the varieties that are planted, some control over the amounts supplied and an advance indication of likely raw material costs which assists in both financial control and production planning (Sections 2.3.4 and 2.7.1). The advantage to the farmer is the security of having a guaranteed market for the crop at a known price, together with any other incentives that may be offered by processors.

However, this type of arrangement can only operate successfully when both processors and farmers honour their side of the agreement. In the author's experience, there have been a number of occasions when these forms of agreement have been tried, but have failed because one party breaks their part of the contract. Typically, this can be farmers who sell part of their crop to traders at each end of the season, when the price is higher than that offered by the processor. The expected volume of crop is not then available to the processor and planned production capacity cannot be achieved, seriously damaging both sales and cashflow. Alternatively, the processor delays payment to farmers, resulting in the need for them to take another loan and greater indebtedness. The processor may also fail to buy the agreed amount of crop and farmers are left to find alternative markets without the option of supplying traders who may refuse to buy it or may offer an insignificant price.

A slightly different approach is that in which a processor takes a greater degree of control over production of the crop and specifies the types of fruit or vegetable to be grown, supplies seeds and other inputs, even including labour. In effect farmers are paid by the processor for the use of their land. Although this involves greater organisational complexity and higher operating costs for the processor, the benefits of an assured supply of raw materials having the correct qualities for processing may outweigh the disadvantages, particularly in situations where the demand for a crop outstrips the supply.

A further development of the approach is for the processor to rent or buy land and set up a separate operation to supply the processing unit. This often happens 'in reverse' when an existing farmer diversifies into processing but retains the farm. In either case the processor hires the labour and supplies all inputs needed to operate the farm. The bulk of the produce supplies the processing unit with any excess being sold in local markets or to traders.

2.6.2 Agreements with Retailers and Other Sellers

Decisions by processors on how to sell their products, and to whom, are part of the marketing strategy for each individual product (Section 2.8.2) and these decisions may therefore be different for each product in a range. For example a jam manufacturer may make one range that is sold to wealthy urban consumers and another that is sold to bakers as an ingredient in doughnuts and cakes. Whatever type of sale is envisaged, it is necessary for processors to understand the market in which they operate and know the way in which products move through the market and gain value. A typical 'route map' for sales of fruit and vegetable products is shown in Figure 49.

As each seller requires a profit of between 10% and 25% for handling, stocking or transporting the foods, it is clear from Figure 49 that the less direct routes from producer to consumer result in substantial increases in the unit cost of the product. Other points to note on the route map are that a lower producer's profit may be necessary when supplying wholesalers that have control over a large part of the market, distributors add a higher percentage than other groups to account for the high transport charges in most developing countries and street traders and kiosk owners usually have the lowest profit of any group.

The simplest form of selling is directly to customers from the processing unit and this method also results in the lowest cost for consumers. Direct selling is common with bakery products, which have a short shelf life and are generally better preferred when straight from the oven, but is less common with fruit and vegetable products. Exceptions include sale of pickles and chutneys from bulk containers into customers' own pots, a small 'factory shop' selling packs of product at the front of the processing unit and sales of fresh, un-pasteurised juices in cafes or tea rooms that are adjacent to the production unit and may also be owned by the processor.

In these cases, provision has to be made in both the design and layout of the premises to accommodate customers and to have staff available to sell the product. In order to maintain control over hygiene, pilferage and health and safety, it is essential that customers are not allowed into the processing area.

Delivery of goods directly to retailers is feasible if the processing unit is located within a reasonably short radius of a sufficient number. However, the costs of distribution are met by the processor and a higher price should be charged to cover this, although it can still be cheaper for retailers than buying from a wholesaler.

Although the above methods are feasible when the production unit is centrally located, in many instances processing is carried out near to a rural area to reduce the cost of transporting raw materials. In this situation, it is more common for either processors to deliver products to one or more wholesalers, or for wholesalers to collect goods from the production unit. The relative advantages and limitations of transporting either raw materials to a production site or alternatively transporting products to a market should be carefully considered when choosing the site of a processing unit during a feasibility study (see also Section 2.5.2). Particular consideration is needed when glass jars or bottles are to be transported on rural roads, because of the potentially high losses of both empty packaging on the way to the unit and filled product on the way to wholesalers or retailers.

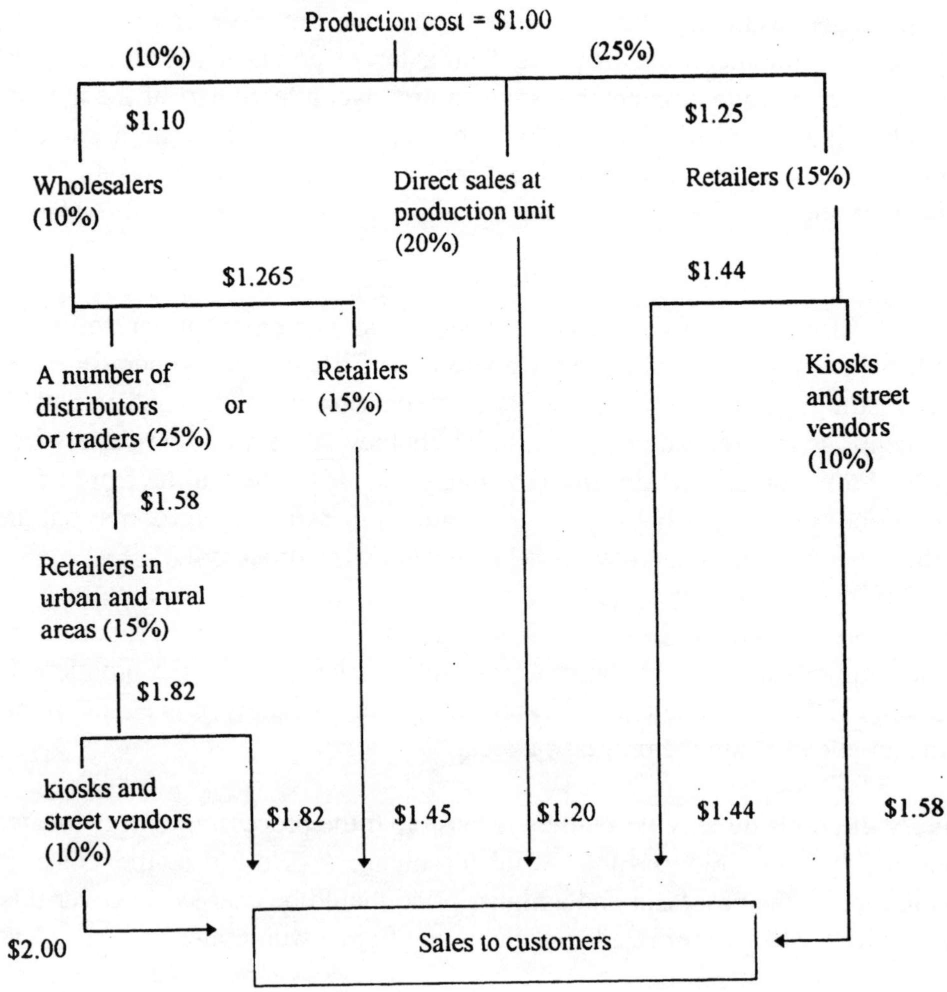

Figure 49. - Examples of sales routes from a processor to the final consumers (percentage figures are profit at each stage, prices are unit sale price to each group)

2.7 Managing Production and Quality Assurance

2.7.1 Managing Production

The design of the processing operation is selected during preparation of the Feasibility Study (Section 2.3.3). Once equipment and facilities are in place, it is then necessary to organise staff for routine daily production. This involves five basic components: production planning, scheduling of inputs, maintenance, staff management and health and safety. These are described below, followed by details of quality assurance procedures needed to prepare quality specifications and to maintain product quality during routine production.

Production planning

The planning techniques that are described in Section 2.3.3 for preparation of feasibility studies are used in a modified way on a daily or weekly basis to plan the activities of the enterprise. The first step is to estimate the likely demand for the products and then use this to plan the amount of production to be undertaken. However, instead of demand estimates arising from market surveys, as in the feasibility studies, the entrepreneur has more up to date information from knowledge of current sales. Among the records kept by a business (Section 2.9), there should be a sales book that details the amount of product sold each day. The book may also record where sales were made and to whom. By adding the daily sales figures to form monthly totals, it is possible to produce a sales graph (Figure 50) that shows the trends in sales for each type of product. From this the owner can then estimate the likely scale of production that will be needed each day or each week to meet the expected sales.

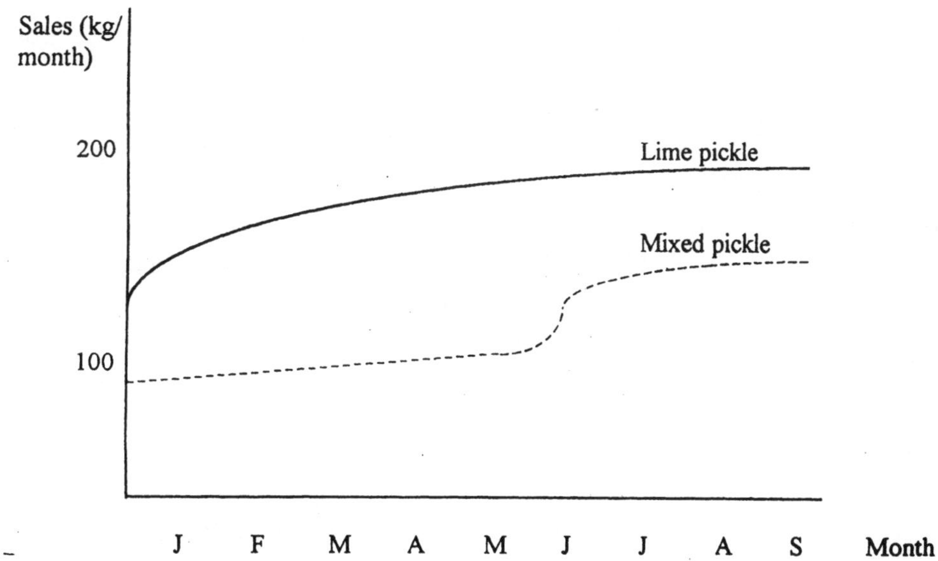

Figure 50. - Example of sales trends for two products

In the example given in Figure 50, sales of lime pickle are predictable, having climbed steadily to 180 kg/month. Sales of mixed pickle were lower at just over 100kg/month before a promotion in May which resulted in an increase to 140 kg/month. In this example, the sales trend is likely to continue at these levels and the volume of production can therefore be planned at 9 kg/day of lime pickle and 7 kg/day of mixed pickle, assuming that there are 20 working days per month. In practice, it may be worthwhile increasing the production rate to 12 and 10 kg/day respectively, to take account of delays in processing through electricity failures, lost working days etc. and to build up a small stock of product. If further promotions are planned it is necessary to increase production beforehand to meet the anticipated increase in demand.

Scheduling inputs

Having decided on the level of production that is needed to meet anticipated sales for the next week or month, it is then necessary to schedule all of the inputs that will be needed to produce the required amount of product. These include not only the components of the products, (raw materials, ingredients, packaging, labels, distribution cartons etc.) but also the number of staff required (especially if they are hired on a daily or weekly basis), cleaning materials, water requirements etc.

With knowledge of the formulation that is used to make each product, the weights of each ingredient and the number of packages that will be required can be calculated. After consulting records of the stocks that are held in store (see Section 2.9.2), orders can then be placed with suppliers to maintain the required levels of inputs.

A common failure of small fruit and vegetable processors is inadequate production planning, so that production ceases midway through a day because for example, one ingredient is used up or the supply of labels or bottle caps is finished. In most developing countries, unless a processor is located near to a large town, it is difficult to quickly replace ingredients or packaging materials and this results in substantially reduced output from the unit. Attention to production planning is therefore crucial to maintain production levels at the planned capacity.

In the author's experience, operation at a small percentage of the planned capacity is one of the most common reasons for failure of an enterprise. This is particularly the case when fixed costs are a relatively large proportion of total costs and the business simply does not produce sufficient product or generate sufficient income to operate above the breakeven point (see Section 2.3.4). In other businesses, the shortfall in production resulting from delays in processing caused by missing inputs, means that the cashflow is inadequate to pay bills, the producer reaches credit limits with suppliers, who eventually refuse to provide inputs.

Because of the difficulties in obtaining ingredients and especially packaging materials in most developing countries, processors are forced to buy larger amounts of stock to protect themselves against intermittent and unreliable supplies. This causes cashflow difficulties both because of the large expenditure to buy the materials and because the cash is tied up for many weeks while stock is waiting to be used.

The alternative of buying inputs more regularly in smaller quantities is often favoured by smaller enterprises to overcome the cashflow problem, but this is not really a solution because items bought in this way are more expensive than buying in bulk and as described above, the risk of loss of production due to intermittent supply is greater. Small businesses are therefore caught in a dilemma of risking a negative cashflow or suffering disruption to production. The problem can be partly addressed by adequate initial financing of the enterprise, perhaps with the facility to take a loan in phases over several months to meet the shortfall in finances caused by periodic negative cashflow (see also Section 2.3.4).

A further difficulty for fruit and vegetable processors, that is less of a problem for other types of enterprise, is the relatively short harvest season for the majority of raw materials. This has three effects: it means that the majority of raw materials must be bought and paid for in a short space of time; where intermediate (or part-processed) products are made, the cash in a business is tied up in part-processed materials for long periods; and when a succession of fruits and vegetables are processed throughout the year, this increases the complexity of production planning because a large number of different ingredients and packaging materials need to be ordered in advance.

Maintenance

Another common reason for lost production is delays caused by equipment breakdowns and waiting for spare parts. Most small scale producers do not have a stock of spare parts for equipment used in their processes, citing the cost as a reason. Equally however, few producers have compared the cost of a stock of spares with the cost of delayed production. This is especially important when equipment has been imported and suppliers of spare parts are not easily contactable, or delivery times are several weeks.

In most enterprises, there are a few items of equipment that are likely to wear out more quickly than others. These include blades in preparation equipment, bearings on motorised mixers, fillers, dryer fans and heating elements in bag sealers. The entrepreneur should therefore identify the specific items of equipment that are likely to fail most often and ensure that a spare component is always kept in stock. Electric motors can be re-wound by small electrical contractors in most urban centres and an arrangement should be made in advance that they will repair equipment as a priority, if the processor guarantees that all such work will be handled by them.

Most small scale processors do not have a programme of planned maintenance of equipment and facilities, preferring to rely on the maxim 'if it is not broken, don't fix it'. There are differences in opinion among engineers over the benefits of a planned maintenance programme. Some believe that it is cheaper to allow equipment to break down and then repair it, whereas others consider that it is cheaper to stop production on a regular basis and replace parts before they wear out. It is likely that the costs and benefits of planned maintenance depend on the speed at which repairs can be done and the value of the spares that have to be held in stock and this is different for every enterprise. As a minimum, managers should monitor the state of equipment and facilities that are likely to wear out and as experience of the rate of failure accumulates over the years, they should buy spare parts or send the machine for servicing at the time the next replacement is anticipated.

Staff management

It is not possible in a book of this type to detail the different features of successful personnel management, but an outline of the principles on which an owner or manager can provide fair and reasonable working conditions for staff is described below.

One aim of a manager should be to ensure that all staff understand the nature of the business and are active in working towards its success. This is particularly important in relation to quality assurance (below) which requires all staff to agree quality management procedures and as individuals, to routinely monitor product quality.

If it is accepted that most people wish to have the following aspects in their job, the manager can arrange work to meet these needs:

- a reasonable wage
- security of employment
- a feeling of belonging to the company
- respect for their skills and knowledge
- good relationships with other staff
- opportunity to develop new capabilities
- reasonable working conditions.

The level of salaries that are paid to processing staff in the majority of small enterprises is usually slightly higher than equivalent work in the Public Sector, to take account of the lack of job security compared to government jobs. However, in many areas where there is substantial unemployment, wage levels are forced down. While any entrepreneur may wish to reduce production costs as much as possible, paying staff below the market rate for a particular job is short-sighted. Trained and experienced process workers are an asset to a small enterprise because they are able to produce products efficiently and to a consistently high quality with minimum supervision. If salaries are too low, workers will seek alternative employment as soon as the opportunity arises, and the expenditure on training and developing their skills will be lost.

The terms and conditions under which staff are employed vary widely in developing countries and in many cases they are employed on a casual basis with no formal contracts or even letters of appointment. In small enterprises, this situation is not likely to change unless governments or other agencies press for changes. However, it is in the entrepreneurs' own interests to create a working environment in which staff members feel that they have security of employment, because they are then more likely to actively work to improve and develop the business. One example of the way that staff feelings of belonging to and sharing in a business could be encouraged is a simple outline of the benefits that they can expect in terms of breaks for meals, amount of pay for sickness or holidays, terms under which absence from work is acceptable (for example for bereavement) etc.

The majority of people wish to have their skills and knowledge recognised and to be able to develop these further in their work. Managers of small enterprises have the opportunity to know the relatively few staff better than in large companies and it is their responsibility to find out what are the skills and aspirations of each worker. Again, it is in the managers' own interests to do this because each worker's skills can then be used most effectively for the benefit of the enterprise. In practice, this may mean allocating specific areas of responsibility, such as record keeping, labelling of products, raw material inspection etc. to those staff that have an aptitude for that type of work. However, it is also necessary to train staff in every aspect of production, regardless of their main area of expertise. When all staff know how to do every job in a production unit, there are opportunities for people to do different work during the day (for example in Figure 33) leading to greater job satisfaction and greater flexibility in job allocation to cover for staff absences.

The owner or manager is responsible for providing reasonable working conditions for employees. This is covered by law in some countries, although it may be infrequently enforced (see also Section 2.4.2). As a minimum, the requirements for hand washing and toilet facilities should be met for both workers' benefits and to maintain hygienic production. Preparation tables should be high enough for staff to work comfortably and where repetitive work is carried out for long periods, as for example in manual packaging and labelling, seats and good lighting should be provided. The owner may also consider providing a rest area with cold water and seating, to prevent workers sitting on stocks of packaging or finished product as the only comfortable place to take a break. These benefits are important to retain experienced staff and contribute to the overall efficiency of production.

Figure 51. - Trained and experienced workers are an asset to a small enterprise

Health and safety

The provision of facilities for staff are important for improved efficiency and staff morale, but the entrepreneur also has a responsibility to staff to provide a safe and healthy working environment. In some countries, this is a legal requirement, but even if legislation does not exist, the consequences of accidents and illness arising from poor working conditions are far greater than any difficulty in ensuring safety.

Most types of fruit and vegetable processing are inherently safer and healthier than some other types of work, such as mining or driving. However, there are dangers in processes that involve heating, particularly when large containers of viscous products such as jam or sauce have to be handled at boiling temperatures. The manager should therefore provide aprons and heat resistant gloves and also train staff to handle such foods safely.

When workers are preparing fruit over several hours, they should be provided with thin gloves to prevent skin damage from fruit acids. This is especially the case with pineapples, which contain an enzyme that attacks skin.

There are also dangers from sharp blades when preparing fruits and vegetables, particularly when motorised cutters or liquidisers are used. Again, it is the responsibility of the owner or manager to ensure that proper training in the correct procedures is given and attention is paid to ensuring that fail-safe devices such as electrical cut-out switches are operational. On larger equipment that is powered by drive belts, there should always be guards in place, staff should be trained to understand safe operating procedures, particularly when cleaning such equipment. The manager should also enforce dress codes to prevent operators from wearing clothes or jewellery that could become entangled in moving equipment.

Dust production is a problem in a few processes and in others the heat and steam produced from boiling pans can produce an unhealthy working environment. The manager should take the necessary steps to extract these from the plant and provide adequate lighting and ventilation to maintain a healthy workplace (see also Section 2.5.3).

2.7.2 Managing Quality Assurance

When a new enterprise is established it is necessary to both standardise the quality of fruit and vegetable products and also ensure that they are safe to eat. The owner should work with the staff to go through each stage of processing, from purchase of raw materials and ingredients to the consumption of the final product, to identify where factors exist that could influence either product quality or safety and to then devise procedures that control those factors. This is known as developing a *Quality Assurance System*.

Implementation of this system starts with processors deciding which focus to address first: improvements to product quality or improvements to product safety. They then examine every stage in the process to find where improvements can be made. If safety is the main reason for doing an analysis, the aim is to draw attention to potential risks and then develop measures to monitor and control the risks so that they do not become a hazard.

Safety of products

Although fruit and vegetable products have a lower risk of food poisoning than for example meat and dairy products, they can still become contaminated with potentially hazardous materials and quality assurance should be an essential component of production planning. In most developing countries, the requirement to produce safe foods in a hygienic way is part of the law and there are serious penalties for those who contravene hygiene and food safety legislation (Section 2.4.2).

The safety of fruit and vegetable products can be assured by implementation of a management method known as the *Hazard Analysis Critical Control Point* (HACCP) system. This is designed to prevent problems from arising, rather than curing them.
In essence, the process of implementing HACCP systems involves the following stages:

- identify potential hazards
- assess the level of risk
- design and implement procedures for
- monitoring and controlling hazards
- apply corrective action in a process
- train all staff in implementation of the procedures
- develop appropriate reporting procedures.

Many small processors may think that development of HACCP systems is not necessary or not possible because it will either be too difficult or too expensive for them.

However, in many developing countries HACCP is no longer a choice but is being demanded by the local Bureau of Standards or by companies that import processed fruits and vegetables. Greater awareness by consumers about food safety and their requirements for improved quality are likely to result in universal implementation of basic HACCP systems in most countries. However, to develop a system, most small scale processors need assistance and advice from professional advisers, including staff at a Bureau of Standards or a university who have experience of the product and the process.

This is especially true when an entrepreneur establishes a system for the first time. This type of assistance is also increasingly seen as a vital service that can be provided by Manufacturers' Associations.

Where an analysis of food safety is required, the stages identified above are implemented as follows:

Identify potential hazards and assess the level of risk: the processing stages are written out as a Process Chart (see Section 2.2 for examples) and ways in which contaminants could enter the food are identified. A selection of different types of contaminant is shown in Table 26.

Types of contaminant	Examples
Microbial	Bacteria, moulds, yeasts, viruses
Biological	Hair, excreta, bone splinters
Chemical	Pesticide residues, detergents
Physical	Bolts from machinery, stones, glass

Table 26. - Some types of contaminants in foods

Up to 95% of customer complaints in countries where these have been monitored, are related to contamination by physical, chemical or biological sources. Microbial contamination is therefore a small part of the risk, but in low-acid foods, the risk of serious food poisoning means that proportionately greater attention is given to this source.

During development of a safety system, emphasis should be placed on

- sources of contamination
- methods of contamination
- effect of the process on levels of contamination
- probability of micro-organisms surviving the process
- and growing in the product.

It is better to first select the most important type of hazard for a particular product and do the study for this. Potentially less important hazards can then be examined later and added to the quality assurance plan. An example of potential hazards and their level of risk in sauce production is shown in Figure 52. Following identification of hazards, the effect of processing conditions on contaminating micro-organisms is then assessed. This should include all parts of the process, from the purchase of raw materials and ingredients to storage and consumption of the final product. Examples of factors that should be examined in a process are the formulation of ingredients, particularly any that are likely to be heavily contaminated, the types of micro-organisms that may contaminate the raw materials, the pH or moisture content of the product and any preservatives that are used. This information is then added to the process chart (Figure 52).

Design and implement monitoring and control procedures: Once the range of potential hazards are identified, control methods can then be developed to prevent contamination. Some parts of a process have greater effect on product safety than others do. Where an error at a particular stage could have an important effect on safety, controls are put in place at these stages. These are known as *Critical Control Points* (or CCPs).

Stages in process	Potential hazards	Level of risk and measures to address risks	CCPs
Harvest			
Wash/sort	Mouldy fruit, contamination with soil, leaves, etc.	High risk: mould contamination could affect flavour and shelf life of product. Moderate risk: extraneous matter such as insects could contaminate product if not removed during inspection. Low risk: cosmetic faults in fruit, other contaminants that would be removed later in the process.	CCPs - no mouldy fruit or insect contamination.
Peel	Seeds and skins not removed		CCP - no peel or seeds in product.
Pulp			
Mix	Contamination of spices with dust, moulds, bacteria or foreign bodies. Correct pH of mixture	High risk: shelf life depends on correct mixture of acid, salt and sugar. Ingredient weight should be checked. Moderate risk: contamination of spices. Remove mouldy items and other contaminants during inspection and washing. High risk: incorrect pH. Product depends on in part on acidity for preservation. Check pH.	CCPs - correct ingredient weights (Figure 17) +/- 5%, correct pH +/- 0.2 units, no mouldy or contaminated spices.
Heat	Insufficient heating	High risk: adequate heating needed to destroy enzymes and contaminating micro-organisms and produce required consistency in product. Check time and temperature of heating.	CCP - heating at 100°C for 20 minutes +/- 5 minutes.
Fill/seal	Faults in glass. Inadequate seal.	High risk: faults in glass could injure consumers. Check by 100% inspection of bottles. High risk: fault seal on cap could allow re-contamination . Check caps are correctly sealed.	CCP - no glass faults. CCP - no faulty cap seals.
Pasteurize		High risk: inadequate pasteurization results in spoilage during storage.	CCP - pasteurize at 88°C +/- 2°C for 20 minutes +/- 1 minute.
Cool			
Label/store			

Figure 52. Process chart showing potential hazards in sauce production

A 'Decision Tree' (Figure 53) can be used to help decide on the CCPs. Target limits and tolerances are decided for each CCP as shown in the example for sauce production in Figure 52.

Train staff to implement procedures: Staff are trained to operate the quality assurance methods. They should also know the limits that are placed on any variation from the specified methods, so that everyone involved in the process understands his or her responsibilities.

Develop reporting procedures: Methods for monitoring quality assurance procedures are designed, together with a plan of what should be done if the tolerances are exceeded. It should be clear who has the authority to make decisions and who is responsible for checking that a corrective action was properly done. This process is not just the responsibility of the owner or manager and it should be developed with the process workers so that everyone is clear about each other's part in the system. The system should be reviewed every year.

Product quality

When a producer wishes to ensure the quality of products, it is necessary to identify where losses in quality are likely to occur and then find methods to control the process and to improve the product. If for example, a problem is due to poor quality raw materials or ingredients, this should be discussed with suppliers and if necessary, the processor should introduce appropriate testing methods with tolerance limits that are agreed with the supplier. If a problem is due to a processing condition, such as the time or temperature of heating, the process control is improved by better staff training, use of thermometers etc.. All changes should be monitored to make sure that they are effective and details of the changes should be recorded in a Production Workbook (Section 2.9.2). Such procedures are intended to control the parts of the process that significantly affect product quality and therefore help the processor to employ staff where they are most effective. Details of quality assurance procedures for each product group are described in the Process Charts in Section 2.2.

Raw materials and ingredients

The main quality factors associated with fruit products are the characteristic flavour and colour of the fruit, the absence of contamination, and in some products, a characteristic texture. However few quality characteristics of fruit products can be measured objectively and fewer still can be measured by machines. Therefore reliance should be placed on subjective assessment by operators and the more operators that examine the raw materials, ingredients, process and product, the greater will be the level of control. The importance of proper staff training and involvement in production are particularly important in fruit processing.

All fruits and vegetables should be processed as soon as possible after harvest to reduce the risk of spoilage before processing. This is particularly important for vegetables to reduce the risk of growth by food poisoning bacteria.

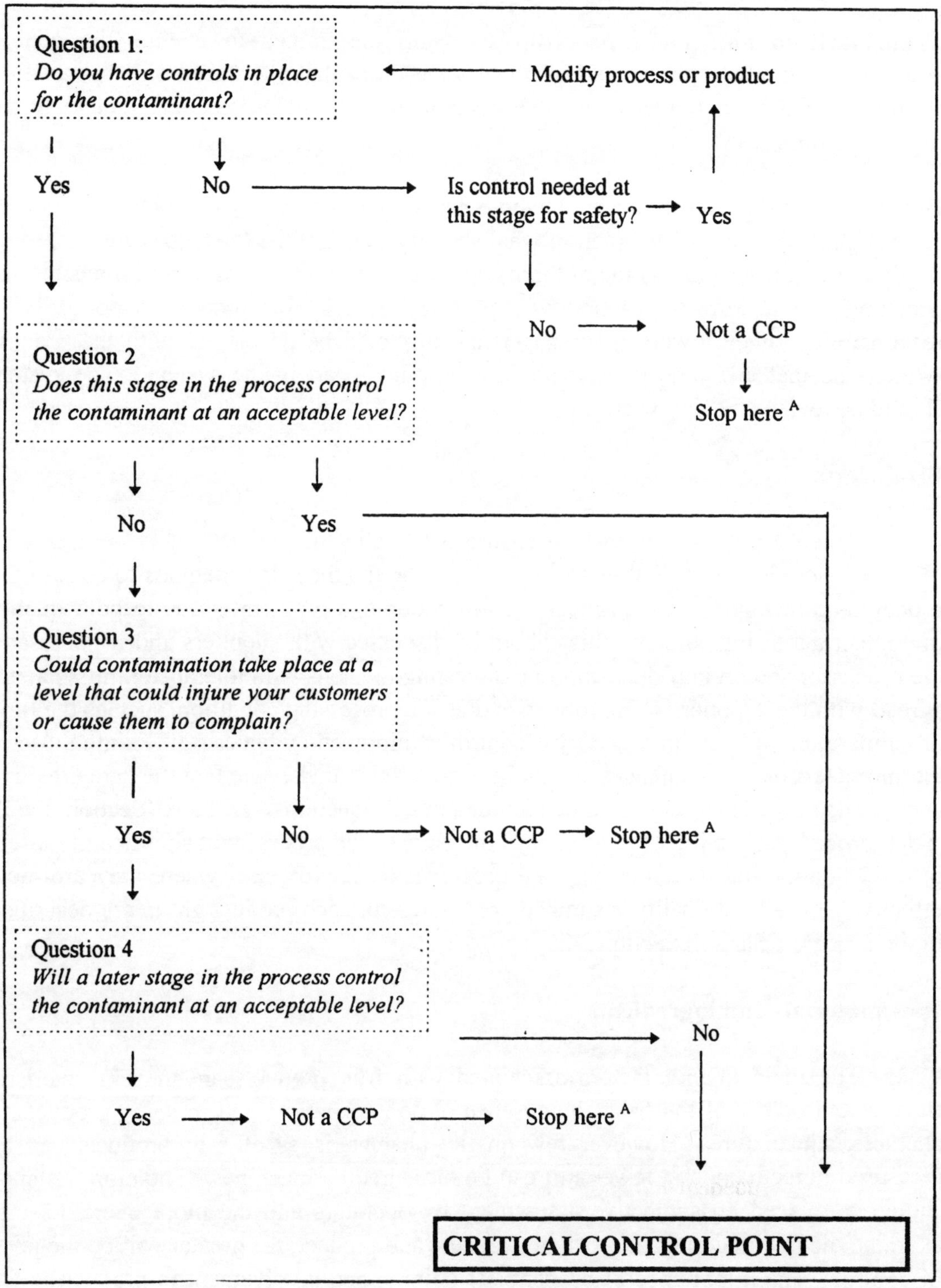

[A] Consider the next stage in the process you are investigating

Figure 53. - Decision Tree' for assessing Critical Control Points
(Adapted from *How to HACCP*, by Dillon, courtesy of MD Associates)

A particular problem facing fruit and vegetable processors in many developing countries is the large number of different varieties of a particular raw material. Not all varieties are suitable for processing and for a processor to be able to make a uniform product, there must be either control over the varieties that are used or a standard system of blending raw materials. Unfortunately this is not always easy to achieve as different varieties of fruits and vegetables are often grown in small quantities by individual farmers.

Orchards or vegetable farms that grow a single variety are unusual and it is a common problem for processors to obtain a sufficient quantity of the required variety, which makes production planning difficult (also Section 2.3.3).

Because most fruits and vegetables mature during a short harvest season, processors must collect and process a large amount of raw materials quickly. In some cases it is possible to part-process and store intermediate products for later production, but there remain specific problems in fruit and vegetable processing that are less evident with other types of processing. For example most fruits and vegetables must be harvested when they are fully mature to give the best flavour and colour in products, but when fully mature many are soft and therefore more susceptible to damage. This damage allows the growth of moulds and yeasts on fruits or rotting bacteria on vegetables. Additionally, damage to a few fruits or vegetables can quickly lead to infection of others and the loss of a whole batch.

Fruits and vegetables should therefore be harvested carefully by cutting them from the tree or plant. With fruits, it is important to leave the stem in place to reduce the risk of infection by moulds and yeasts through an open stem hole.

Bad practices at harvest cause many problems for the processor later on, but the processor often has no control over harvesting methods and the farmers do not understand the processors' requirements. In this situation, it is therefore advantageous for the processor to work with farmers to improve the quality of raw materials. Examples of ways that processors can do this are as follows:

- handlers should be asked to cut their fingernails to prevent them puncturing fruits
- in tropical climates, fruits and vegetables should be cooled after harvest to remove some of the 'field heat' and stored in a cool place or covered with wet sacks
- any damaged pieces should be removed from the bulk as they will lead to rapid spoilage of surrounding foods before processing starts
- fruits and vegetables should not be thrown into piles. They should be filled into crates that are small enough to be carried and not dragged along the ground

- crates should not be over-filled as this crushes the food if boxes are stacked. Ideally, foods should be packed into stackable crates which prevent crushing.

An important aspect of raw material supply is the relationship between growers and the processor and each must have trust and confidence in the other for long term honest dealings.

The agents who normally buy crops from farmers sell to different wholesalers, and some wholesalers buy lower quality for cheaper markets. This means that the transporter is always able to sell a complete load, regardless of the quality. As a result the transporter has no particular interest in safeguarding the quality of the crop. Commercial pressures to carry a maximum load on each journey also mean that the transporter is more likely to pile fruits and vegetables onto a truck, rather than using crates that take up extra space and reduce the value of the load. The agents also carry consumer goods on the return journey to rural areas and are unwilling to return crates to farmers unless they are paid to do so. In the author's experience they are even unwilling to carry collapsible crates which take up less space, for the same reasons. Processors thus have a difficult problem in controlling the quality of raw materials if they do not collect fruit and vegetables themselves from farmers. Contracted farming and management of raw material supply is discussed in more detail in Section 2.6.1.

The first inspection of raw materials may therefore take place as they arrive at the processing unit. The inspection should check that the fruits or vegetables are suitable for processing and reject those that are not. This normally includes a check on the following characteristics (see also Section 2.2 for individual processes):

- maturity (over-ripe or under-ripe)
- colour
- size or shape (for some products)
- visible mould or rots
- serious bruising or cuts
- presence of soil, large amounts of leaves or other materials.

The percentage of rejects should be monitored as this is an important factor in calculating the true cost of useable raw material (see Section 2.3.3)

At this stage in processing, careful inspection by properly trained staff is an important method of maintaining product quality and saving time and money later in the process. It should be remembered that *poor quality raw materials produce poor quality final products.* It is not possible to improve the quality of raw materials by processing them. Sorting out substandard materials before money is spent processing them is therefore one of the most cost effective methods of ensuring a uniformly high quality in the final product. Similarly, other ingredients should be checked to make sure that they are the correct type and are not contaminated or adulterated.

Processing

After initial inspection, fruits and vegetables are washed in potable water, which is chlorinated if necessary (Section 2.5.3). It is important that the process staff are trained to remove any pieces that are rotten as these would quickly contaminate the wash-water and infect good quality raw materials. They should also remove all leaves, insects and other wastes that could contaminate the final product.

The next stages in processing involve preparing fruits and vegetables by peeling, slicing, pulping or filtering. Quality checks during preparation stages are to ensure that all peel is removed, the yield of useable material is calculated (Section 2.3.3) and that slicing produces uniform sized pieces for products such as banana chips and shreds for marmalade. Any over-sized pieces should be re-sliced. Fruit pulps are filtered using nylon or muslin bags or special juice filters, when a clear product such as squash, jelly or juice is required. Quality control measures during this part of a process are to ensure that bags are properly washed and boiled for at least 10 minutes before each use and that the juice has the required clarity.

Strict hygiene in the processing room and by operators is needed to reduce the risk of both food spoilage and food poisoning. This must include proper cleaning of knives, drying equipment and processing rooms, washing hands, and removal of waste foods as they are produced (Appendix I).

The correct formulation of a batch of ingredients for subsequent processing is critical to both the quality of the final product and the financial viability of the operation. Good control at this stage enables a uniform product to be made in every batch and saves money by not wasting expensive ingredients. Additionally, any mistakes that are made at this stage cannot be easily corrected later and may result in having to throw away a whole batch.

The staff responsible for batch formulation should therefore be given thorough training and a management system that records batch numbers and amounts and types of ingredients used should be put in place (see also Section 2.9.2). When producing dried fruits the amount of residual sulphur dioxide in a product is controlled by law in many countries and sodium metabisulphite for a sulphite dip or of sulphur for burning in the cabinet should each be carefully measured out.

Equipment required for batch formulations includes good quality scales or calibrated cups, spoons or jugs to measure out ingredients and ensure that the same amount is added in every batch. Sugar concentrations in jams, sauces, syrups etc. can be checked using a refractometer (Figure 12). Although this equipment is relatively expensive, it gives an accurate measurement of sugar concentration which is essential for the correct preservation and anticipated shelf life of many products. The reading is recorded as degrees Brix which corresponds to % sugar. The concentration of sugar in syrups can be measured more cheaply using a hydrometer with a scale reading %

sucrose. The syrups should be tested at 20°C, which is the reference temperature for the hydrometer.

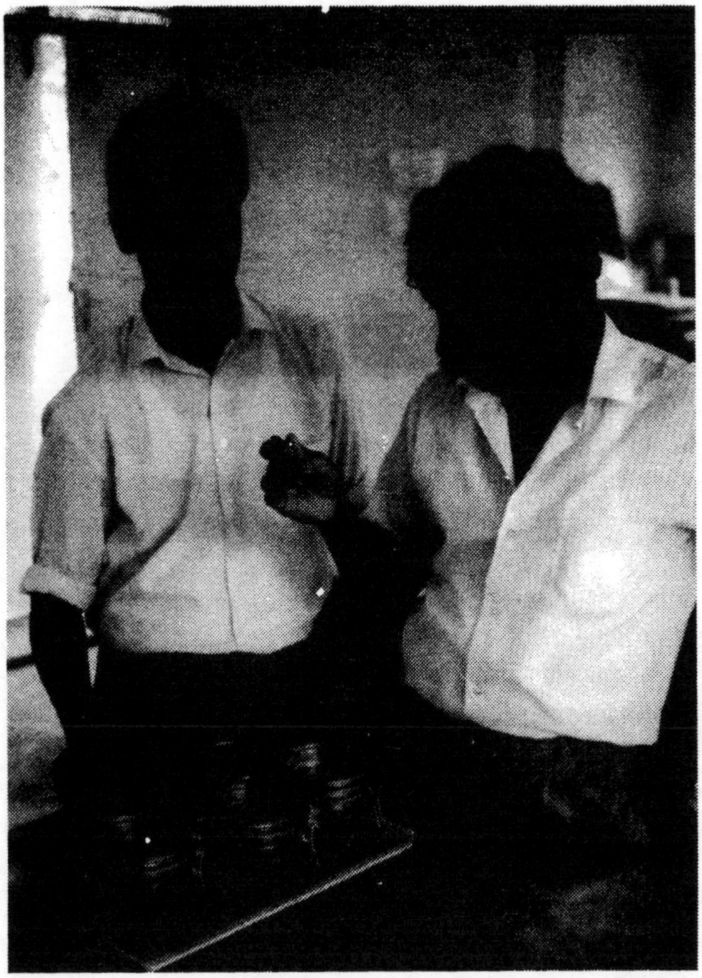

Figure 54. - Correct formulation of ingredients is an essential feature of quality control

As the sugar content of jams and fruit cheeses increases, so does the temperature of boiling. The sugar concentration can therefore be estimated by measuring the temperature, using a special thermometer that reads up to 150°C. However, the boiling temperature is also affected by the amount of invert sugar in the mixture and staff should have experience of making the product before using temperature alone to control the process. The boiling point also changes with height above sea level and in mountainous regions, producers should first check the boiling point of water and make the necessary corrections. With experience, staff can also estimate the solids content of preserves by cooling a sample of the boiling mixture and noting the texture to see if a firm gel forms.

In some formulations it is necessary to check the pH of a product or the amount of acid that is present. pH is a scale from acidity (pH 1-6), through neutrality (pH 7) to alkalinity (pH 8-14). It can be measured by dipping a piece of pH paper into a sample of liquid food and comparing the colour change with a chart supplied with the paper. For greater accuracy a pH meter should be used.

It should be noted that pH measurement does not tell you the <u>amount</u> of acid present. This is especially important in pickling, where preservation is achieved by the correct combination of acids, salt and sugar. To measure the amount of acid in a product (such as citric acid or acetic acid), a 10g sample of food is mixed with 90ml of distilled water and 0.3ml of indicator solution such as phenolphthalein. This is then titrated with 0.1M sodium hydroxide until the pink colour does not change. The amount of acid is calculated using the formula:

% acid = number of ml of sodium hydroxide x one of the conversion factors below:

acetic acid (vinegar) = 0.060, citric acid = 0.064, tartaric acid = 0.075, lactic acid = 0.090. It is necessary to know type of acid in the food before selecting the conversion factor.

The salt concentration in pickling brines can be measured using a special hydrometer (Figure 55). Although other methods, such a salt refractometer and titration exist, they are too expensive or complex for most small scale processors. A sample of brine at 20°C is filled into a large clear glass or plastic cylinder and the hydrometer is placed into the liquid. When it has stopped moving, the scale is read at the surface of the liquid and the reading is converted to % salt using a conversion table supplied with the hydrometer.

Chutneys and pickles rely for preservation on the correct balance of acids, salt and sugar in the final product and they are often not pasteurised. Strict control over hygiene is essential as insects can contaminate the pickle with large numbers of moulds and yeasts during pickling and these can spoil the product during storage. The pickle should therefore be protected from insects and covered at all times to stop dust and other contamination.

Products such as sauces, preserves, drinks and bottled fruits are each heated during processing and the time and temperature of heating is an important quality check. Over-heating causes lowered quality by loss of texture, colour or flavours, whereas under-heating allows enzymes and contaminating micro-organisms to survive and later spoil a product during storage. It is therefore essential that an adequate temperature and time of heating are carefully controlled and operators are trained to ensure that these conditions are maintained for every batch of product. The equipment required for process control includes a clock, a thermometer and for concentrated products such as jams and fruit cheeses, a refractometer to check the final solids concentration.

The shelf life of dried foods depends mostly on the *equilibrium relative humidity* of the product under the expected storage conditions (Section 2.2.3). This is usually found by measuring the moisture content of the product as described below, but because the relationship between moisture content and humidity varies with different foods, it is necessary to conduct trials to find the highest moisture content at which the food will be stable. This involves taking samples at different times during drying, packaging them and after storage for three to four weeks, checking them for spoilage. The ones that have not gone mouldy are then checked to find the moisture content and this becomes the target level for subsequent production. During drying, the air temperature and drying time should be carefully controlled to ensure that fruits and vegetables are fully dried to the required moisture content.

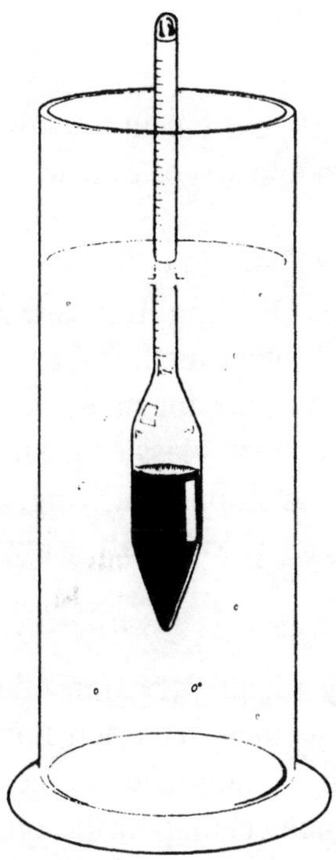

Figure 55. - A brine hydrometer

The moisture content can be found by carefully drying a known weight of finely chopped food in an oven at 100°C until it does not lose any more weight. The final weight is noted and the '% solids' is calculated using the following formula:

$$\% \text{ solids} = \frac{\textbf{final weight of sample}}{\textbf{initial weight of sample}} \times 100$$

% moisture content is then 100 - (% solids)

The methods described above for process control are each relatively simple and have sufficient accuracy for routine use. They do not need sophisticated or expensive equipment or high levels of skill and they are sufficiently inexpensive to be used routinely by small scale processors. However, many of the methods are comparative and the results can only be compared with other results obtained by the same method.

This is acceptable for routine process control, provided that careful attention is paid to ensuring that exactly the same procedure is followed each time.

The time spent on process control should be greater than that spent on testing the final product, because it is better to have control of the process and prevent mistakes from occurring rather than trying to correct a badly made product (prevention is better than cure). This is the basis of the HACCP approach and quality management systems should reward operators for reporting and/or correcting faults in a process as they occur.

Packaging, storage and distribution

Although fruits and vegetables are stabilised by processing, for many their long term preservation depends on the type of package that is used and the temperature and humidity in which packages are stored. For these reasons, it is important that packaging, distribution and storage are included in a processor's quality assurance schedule. Details of the methods of manufacture of packaging materials and potential faults that should be tested for are included in publications in the Bibliography, and a summary of the main quality assurance checks on packaging materials is given below.

The risk of glass splinters in a product which would cause serious harm to consumers, means that bottles and jars are subjected to more rigorous quality checks than other types of packaging. It is essential that all glass containers are checked to ensure that there are no glass splinters or cracks, bubbles in the glass or strings of glass across the interior. Staff who check bottles or jars should be fully trained in the faults to look for and they should only work at inspection for 30-60 minutes at a time to maintain their concentration.

Because of the way in which they are made, the dimensions of glass jars and bottles are also more variable than other types of packaging. It is therefore important to check that a container has the expected capacity, that the neck is properly formed and will allow the lid to fit and that it stands vertically to prevent it breaking in a capping machine. It is also necessary to check the weights of a number of empty jars or bottles to find the heaviest.

This is then used in checkweighing (the checkweight is the weight of the heaviest container plus the net weight of product). If jars or bottles are re-used, they may have contained poisonous materials such as pesticides. They should be thoroughly washed and inspected by smelling them to ensure that there are no residues.

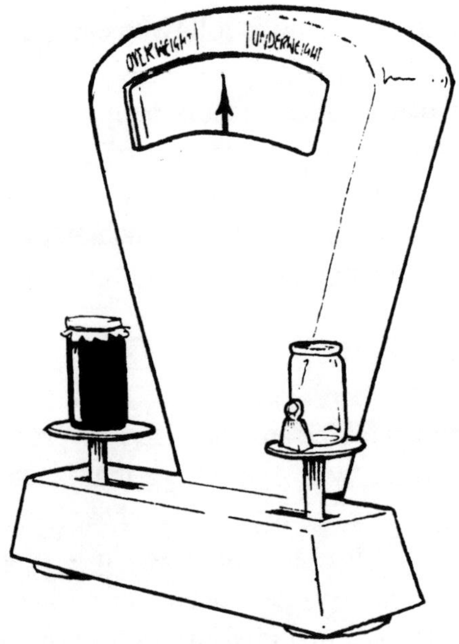

Figure 56. - Checkweighing scales

Filling products into containers is an important control point in a process because in most countries it is an offence to sell an under-weight product and over-filling means that a producer is giving product away. All products should therefore be carefully filled to ensure that the fill-weight is the same as the net weight described on the label (see Sections 2.4.2 and 2.8.3 for details of labelling requirements). A random sample of packages should be checked to ensure the correct net weight using a checkweighing scale (Figure 56). Other information that should be checked at this stage includes whether the label matches the actual product in a pack, that the sell-by date on the pack is correct and that batch code numbers are correct. All data should be recorded in a Production Logbook (Section 2.9.2).

The capacity of a jar or bottle can be checked by weighing a dried container, filling it with distilled water and re-weighing it. The difference in weight is equivalent to the capacity in ml and should be large enough to contain the net weight declared on the label. When a product is filled, there should be a space between the surface of the product and the underside of the lid. This *headspace* not only allows a partial vacuum to form when hot-filled products cool, but a consistent level of filling is an attractive marketing factor. A simple gauge for checking that product has been filled to the correct level can be made locally (Figure 57). It is placed on the rim of the jar and the level of product can be read where it touches one of the prongs.

Other routine checks for glass containers are the diameter at the neck and body using *go/no-go rings* (Figure 58) and that containers are round and not oval. Rings are placed around the neck or body to show whether the neck diameter is too large or too small for the lid, whether the neck is circular, to check the outside diameter of the container and whether it is oval or round.

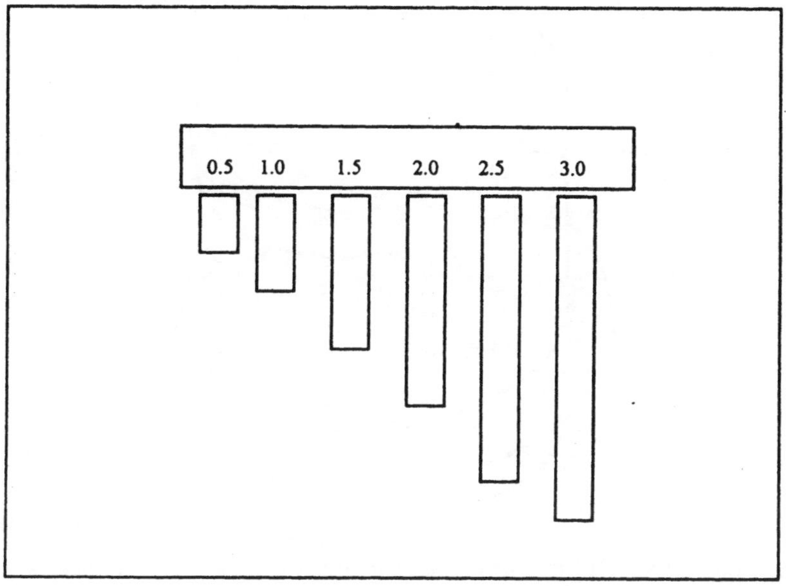

Figure 57. - Headspace gauge

In plastic bags and films, typical faults include 1) incorrect printing, 2) smelling of the odour of solvents used in their manufacture, 3) layers of film on a roll sticking together, 4) poor seal strength, 5) curl, in which a film curls up rather than laying flat and 6) incorrect thickness (known as 'yield'). The last can be measured by cutting 10 squares of film, each 10cm by 10cm and carefully weighing them. The result (in grams/square metre) is then checked against the suppliers' specification.

Value is added to raw materials at each stage of processing and by the time it is packaged it has gained most of its final value. Any losses of packaged product are therefore the most serious, resulting in the greatest loss of money to the processor. Great care should therefore be taken in handling and storing packaged foods and they should be stored in boxes on pallets or shelves to keep them off the floor. The storeroom should be cool and dark with a good ventilation to maintain a flow of air and with protection against insects and rodents. Storerooms for ingredients and packaging materials should have similar protection (also Section 2.5.3)

Quality management systems should also be developed to monitor the types and amounts of products, ingredients and packaging materials that are in the storeroom and the time that they remain in storage. Records should be kept by storekeepers to show which materials are transferred into and out of the storeroom and when they were used (see Section 2.9.2). Similarly, the control of product quality does not finish when the product leaves the processing unit and manufacturers should monitor and control the distribution methods to retailers and discuss with them the best ways of storing and displaying the products.

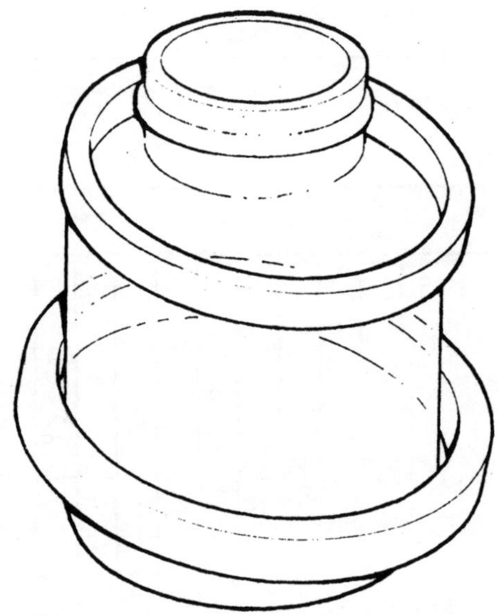

Figure 58. - Go/no-go rings for measuring glass containers

Hygiene and sanitation

Together, a manager and processing staff should apply the HACCP approach to identify all areas of potential hazard in the production of a food and then develop a cleaning plan and personal hygiene rules that ensure safe preparation of the product. The manager should monitor the plan and make sure that all staff are trained and know their own responsibilities. Similarly, it is important that staff are not penalised for having an infection, otherwise they will hide a problem in order to be paid. If staff report a stomach illness or skin infection, they should be transferred to other jobs that do not put them in direct contact with the product. The manager should also provide proper cleaning materials and equipment and allow adequate time for cleaning machinery and processing areas after production has finished.

Cleaning schedules should be drawn up when specific areas of hazard have been identified in a process or in the building. All areas need attention but some carry a greater risk than others. Each worker should know their cleaning responsibilities within a cleaning plan and the manager should take overall responsibility to ensure that cleaning is done to the correct standard and that a cleaning schedule record is maintained. A summary of the quality assurance procedures for a typical fruit and vegetable process is shown in Table 27 and details for individual product groups are given in the Process Charts in Section 2.2.

Stage	Process Activity	Control Point
Raw material harvest	Liaison with farmers, pick fruit and load it into crates or purchase from markets.	Specifications of fruit quality required. Training of pickers and handlers to minimise damage to fruit. Use of correctly designed boxes. Rejection of damaged or rotten fruit and vegetables.
Raw material transport	Transport in crates to processing unit.	Control of fruit temperature by use of water, shade or covers. Correct stacking and handling to minimise damage. Reduce delays and minimise journey time.
Raw material inspection and preparation	Record amount and quality of fruit received. Sort fruit, wash and peel/slice as required for the specific product.	Setting of acceptable standards for incoming fruit. Training in correct sorting, preparation and recording procedures and management to ensure procedures are implemented. Accurate slicing to required sized pieces. Operator hygiene and plant hygiene. Water chlorination. Regular disposal of waste.
Ingredient formulation/batch preparation	Weigh and mix ingredients.	Training in accurate weighing and keeping records of ingredients used.
Processing	Part-processing if fruits are to be stored for later use. Heating, drying, pickling etc. to make the required product.	Preparation of processing schedules and training of operators to ensure: - control of temperature and time of heating or drying, - correct amounts of ingredients added at the correct time in the process. Establish standards for operator hygiene and schedules for cleaning of equipment and processing room.
Packaging	Fill product into packages, seal and label. Pack into distribution boxes.	Establish specifications for package quality (especially glass containers), labels and fill-weights. Implement inspection, check-weighing and recording procedures.

Table 27. - Summary of quality assurance procedures for fruit processing (From: *Quality Assurance for Small-Scale Rural Food Industries*, courtesy of FAO)

2.8 Marketing

Perhaps the most common problem faced by small scale producers is their inability to effectively sell their products. There are numerous examples of producers who are able to make high quality products at a competitive price, but have little experience or skill in finding people who are willing to buy them. Market research and the development of an effective marketing strategy are therefore essential components of establishing and running a small fruit and vegetable processing enterprise. In Section 2.3.2, the procedures that are used to conduct market surveys are described. This is the first stage in identifying different markets and is a necessary step in developing a marketing strategy. Subsequent steps in developing a marketing strategy are described in more detail below.

2.8.1 Identification of Markets

There are always a number of different markets into which fruit and vegetable processors can sell their products. Within each market there are also a number of *market segments* or sub-divisions, that can be specifically targeted by a producer. It is very important to decide at an early stage in establishing a business, what type(s) of market does a processor wish to target but also which particular segments within each. These decisions should be evaluated (and if necessary changed) at regular intervals.

It is true that in many developing countries the total market for some types of processed fruits and vegetables is small and selling to customers in one particular market segment may not be sufficient to exceed the break-even point for a small enterprise. However, the process of identifying different market segments helps the entrepreneur to focus on how the business will operate and what types of promotion, distribution and selling should be used.

Examples of different types of markets for processed fruits and vegetables are shown in Table 28, together with examples of different segments within each. It should be stressed that this list is not comprehensive and that other potential markets and market segments may exist in specific locations. The entrepreneur should therefore find out the types of market that exist locally and determine whether they may be suitable to supply.

Market Segments

This is the term given to different identifiable groups of customers. Market segments are described by different income levels but examples of other segments include:

- those based on age (e.g. foods that are mostly eaten by children, such as sweets or weaning foods)

- those based on sex (e.g. foods that may be mostly eaten by men, such as snacks that are taken in bars)
- those that are based on religious beliefs (e.g. special foods for festivals)
- those that are eaten mostly by office workers at lunchtime etc.

Within each broad market type, there are a number of segments that may have different needs for particular types of fruit and vegetable products. If a particular segment is targeted by a producer, this is known as selecting a *market niche* and a product that is sold to a single market segment is known as a *niche product*. An aspiring entrepreneur should carefully consider which types of people are likely to buy a new product and then devise promotion and sales methods that suit the groups that are selected. A checklist of market information that should be sought may include the following items:

- who will be your customers (businesses, institutions, private individuals)?
- where are your customers (urban, rural, which towns, nearby to production site)?
- what are the average income levels of your intended customers?
- who are the important competitors, how many are there?
- where are your competitors?
- what are their apparent successes and weaknesses?
- how will your product be better?
- who will sell your product and where are the sellers located?
- how will your product be distributed?
- how will your product be packaged?
- what promotion or advertising do you intend to do?
- what further information do you need to obtain?

For example, in the urban domestic market in there may be different segments based on income levels, on gender or age, on eating habits such as vegetarianism, on types of work that people do or on particular areas of concern such as 'healthy eating'. Similarly in the institutional markets the segments may include food for children in rural schools, foods that are used in meals for patients in district hospitals or for the soldiers in military barracks.

The importance of identifying these different segments is three-fold: first it is possible to tailor the product quality characteristics to those that a particular group of customers say they prefer; secondly the promotion of a product can be designed to target a particular segment and thirdly the distribution and sales outlets can be chosen to target where people in the particular segment usually buy their food.

Taking fruit bars as an example, these can be made to compete with alternative products such as sweets, the consumers in both rural and urban areas are likely to be children but the customers will differ depending on the location. In rural areas where there may be less disposable income, mothers buy an individual sweet for their children from village shops or at weekly markets as a reward or for a special occasion. A father may buy sweets as a special treat when he returns home from a period away from the household.

In urban areas families may have more disposable income, a higher level of knowledge about dental problems caused by excessive sweet consumption and a desire to eat more 'healthy' foods. These mothers may therefore prefer to buy a fruit bar as they perceive that it will be better for their children's health than more traditional sugar based sweets. In cities they may be bought by mothers from supermarkets in bags that contain a larger amount, which is given to children over a period of time.

Alternatively, people may give money to children to buy their own sweets from local kiosks. In this example there are therefore a number of market segments that a producer may wish to target:

- rural mothers who buy individual sweets from village shops or weekly markets
- rural fathers who buy individual sweets from kiosks at bus-stops or village shops
- urban mothers who buy packets of sweets from supermarkets or local shops
- urban mothers who prefer to buy fruit based sweets instead of sugar based sweets
- urban children who buy individual sweets from local kiosks.

The fruit bar producer may therefore wish to address the market segments of concerned mothers in urban areas as well increasing the promotion in village shops and kiosks. The results of a market survey (Table 14) can be used to determine the size of different market segments that could be targeted. In the example from an East African country, shown in Table 28, assumptions are made about the percentage of people in each segment who would buy the processed food (a fruit based snack-food) and it can be seen that the two main market segments are rural poor and expatriates/tourists.

Although only a small proportion of poor rural people buy snacks each week, the large numbers involved mean that the market size is also large. In the other category, the higher disposable incomes of tourists and the larger percentage that are expected to buy fruit based snacks or have them available in hotels makes this an important potential market segment.

Distribution and promotion

Each market segment may require different types of distribution and promotion. In the rural market, distribution is via wholesalers who transport the product to a number of rural towns, together with all other goods that are sold in village shops. The village shop owners then visit the towns to buy stock using public transport. The product therefore increases in value and in price each time it is handled by a distributor or trader and a price mark-up of between 10% and 25% can be expected at each stage (see Figure 49). The tourists buy snacks from a variety of sources: from kiosks and restaurants along the tourist routes, from supermarkets in towns and at hotels. Depending on the area that a producer wishes to cover and the number of such sales outlets, it may be possible to supply them directly and avoid price increases by traders. The types of promotion that are available to producers are as follows:

- newspapers
- radio and television
- signboards, posters and leaflets
- personal contacts
- special promotions
- free samples in retailers' shops.

In the examples given above, the types of promotion are different for each market segment. Rural customers are unlikely to have access to television, but may have access to a radio or to newspapers. However, posters or signboards in villages and special promotions in retailers' shops are likely to reach more people. Tourists are unlikely to use radio, TV or newspapers, but may see signboards or kiosk advertisements and buy the product along tourist routes. However, personal contacts with hotel owners and promotions in supermarkets may be more effective.

2.8.2 Developing a Marketing Strategy

From the above examples, it can be seen that a processor should identify as precisely as possible who the main consumers will be, where they are located and how they buy their foods. When this information is added to that about the quality and price that consumers expect (Section 2.3.2), the result is known as the *marketing mix* which is often described as the '4Ps' - Product, Place, Promotion and Price. Some aspects of the marketing mix are described in Figure 59.

Using this information, producers can then refine their product to meet customers' needs and develop a strategy to market their products to the particular segments that they believe will provide the greatest sales. This involves creating a product with the characteristics of flavour, size, appearance etc. required by the customers, developing a suitable and attractive package (see below), negotiating with wholesalers, retailers, distributors, hotel and restaurant owners designing and distributing promotional materials and finally producing and supplying a uniform quality product in the amounts required.

Marketing is therefore putting systems in place that will both make consumers believe that they are buying something special that meets their needs and also supplying

Type of consumer	Estimated number of consumers, assuming different %'s of total actually the processed food						
	0.1% buy	0.5% buy	1% buy	5% buy	10% buy	30% buy	Total**
Domestic:							
Urban wealthy				321			6,424
Rural wealthy		198					39,522
Urban not wealthy	61						61,250
Rural not wealthy	758						758,600
Tourists/expatriates						1155	3,500
Customers in 6 restaurants/snackbars			5				520
Ferry customers			60				1200
Institutional:*							
Hospital (x1)							0
Schools (x17)							0
Prison (x1)							0
Army barracks (x1)							0
Other food processing companies*							0
Export: Urban wealthy across border	1						1300
Customers of International 'Fair Trading" companies	25						25,000
International import Agents*							0

* Survey findings revealed that these potential customers were unwilling to buy the product

** Totals obtained from census data, socio-economic survey data and information from interviews

Table 28. - Estimated sizes of different market segments for fruit bars

the right amount of product when the customer wants to buy it. Customers' perceptions are not just about price and quality, but may also include status, enjoyment, attractiveness, convenience, health or nutrition. Producers should decide which factors are special for their product and emphasise these in their promotion.

Product	**Place**
Better quality Better appearance More attractive packaging Clearer labels More nutritious More varieties Different colours Better flavour Available in required amounts	Longer opening hours Better decoration Cleaner environment Popular location Delivery service Fast and friendly service Good range of stock Ease of supply
Promotion	**Price**
Advertising Free samples Competitions and shows Articles in newspapers Special promotions In-shop displays	Lower prices Discounts for higher quantities Special offers Credit facilities

Figure 59. - Examples of components of the marketing mix

It should be noted that the development of a marketing strategy is not a single exercise that is done when a business starts. The strategy should be continually monitored, to see if planned sales are taking place and the expected customers are actually buying the product. The strategy should be constantly reviewed to improve it or even to change it completely. In this context, the actions of competitors are critical (see Section 2.3.2). It is most unlikely that other producers will do nothing when a new product is promoted. They are more likely to react by offering loyalty bonuses to retailers who continue to promote their products, to introduce their own special offers and increase the amount of promotion that they do. A new producer should therefore be constantly aware of the feedback from customers and retailers, the changes that competitors make and any customer complaints that are received.

2.8.3 Packaging and Brand Image

At an early stage in the development of a new business, entrepreneurs should decide on the symbol or image that will be used to identify their products and makes them recognisable among those of competitors. This 'logo' is used on all products in a producer's range and helps to develop a *brand image*. The label on a package is the first point of contact between a customer and the producer and it should therefore be considered as part of the marketing strategy. If first time buyers are attracted by the label and enjoy the product, they will continue to buy the same brand and develop a loyalty to it. These repeat buyers are the type of customer that is required to build up sales of a product.

The label not only gives customers information, such as what type of product it is and how it is used, but the design and the quality of printing also suggests to the customer an image of the product. This can be one of high quality, exciting taste or a reliable company, but a poorly produced label can also suggest low quality food, lack of care in its production or a cheap product that is only eaten by people who cannot afford anything better. When products are displayed in retail stores alongside those of competitors, including imported brands, the package and particularly the label has to compare favourably with the others before customers will choose it.

The design of a label and the quality of the paper or other materials that are used is therefore of critical importance in promoting the product. In general a simple, uncluttered image on the label is better than a complex design. The brand name or the name of the company should stand out clearly and if pictures are used, they should be an accurate representation of the product or its main raw material. Examples of good label design and promotion are shown in Figure 60a and b and in Figure 61.

Colour can be used to produce either a realistic picture (full colour printing) or blocks of one or two bold colours to emphasise a particular feature. Care is needed when choosing colours as they are culturally very significant and have a direct effect on peoples' perceptions of the product. For example in many societies, white is associated with death, whereas in others, it is red or black. In some areas, browns, ochre and greens are associated with 'nature' or natural unprocessed products, with an image of health and good quality. In others, bright oranges and yellows can either mean excitement or cheap, low quality products.

In some countries there are legal requirements on the design of the label and the information that is included (Section 2.4.2). As a minimum in most countries, the following information should be clearly visible:

- name and address of the producer
  name of the product
- list of ingredients (in descending order of weight)
- net weight of product in the package

- a 'use-by' or 'sell-by' date.

An example of a label that contains sufficient information is shown in Figure 61. In addition, the producer may wish to include:

- instructions for preparing the product
- storage information or instructions on storage after opening
- examples of recipes in which the product can be used
- an 'e-number' if export to Europe is contemplated
- a bar code.

In view of the importance of labels, producers should pay the highest price that they can afford to obtain the best possible quality. Professional designers or graphic artists may be located at universities, art schools or in commercial agencies and these should be employed to produce a range of ideas. These can then be discussed with the Bureau of Standards and then a printer to obtain quotations before a final decision is made. Most printers require a print run of several thousand labels and great care should be taken to check the design for errors before printing, as these would be very costly and time consuming to correct during production. The subject of labelling and package choice and design is complex and is described in more detail in the publications in the Bibliography.

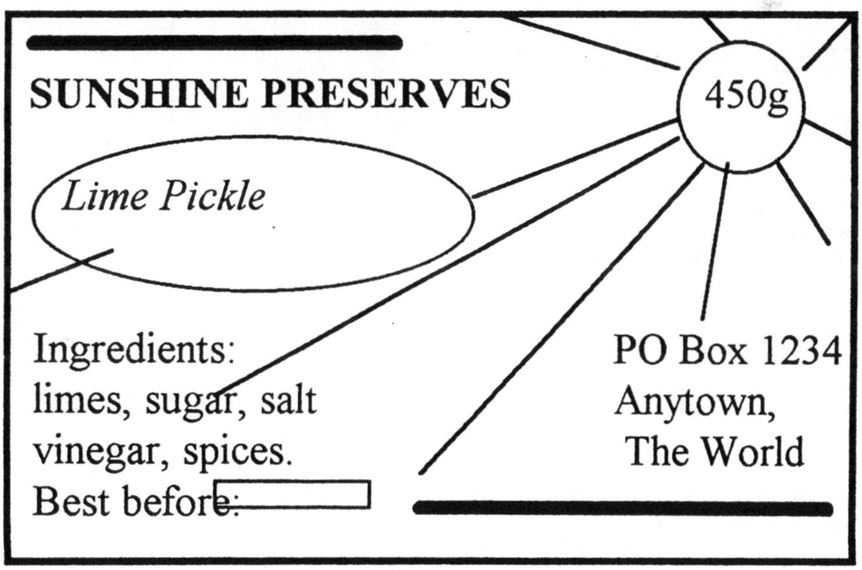

Figure 60a - Well-designed label

Figure 60b - Well-designed label

Figure 61. - -Promotion of products using a poster

2.9 Record Keeping

There are three sets of basic records that should be kept by the owner of a small fruit and vegetable processing unit: financial records, those that relate to the production of the products and sales records. The uses of these records are inter-related and are described in more detail in Sections 2.3 and 2.7. In this Section, the format of the records and the likely ways in which information will be obtained are summarised.

As with all other inputs to a business, keeping records is an investment of time and money and the benefits must outweigh the costs. There is no point in recording information for its own sake and records must be used if they are to have any value. This means that the owner or manager must understand why the information is collected and what it can be used for. Similarly, the time and effort spent in keeping records must be related to the scale and profitability of the business. While it is true that some successful entrepreneurs keep all of the information in their head and do not keep records, no-one else can help run the business during times of illness or absence. Some examples of the value and costs of keeping records are shown below:

Value of record keeping:

- detailed knowledge about the operaticnof the business
- identification of trends
- accurate control over finances and prodouct quality
- identificationo f individual costs to allow changes to a product or process to optimise profits
- keeping track of money owed to the business
- evidence for tax authorities (may be a legal requirement)
- factual basis for product pricing or salary levels
- knowledge and avoidance of theft.

Costs of record keeping:

- time spent learning how to keep records or training staff time spent writing them
- cost of materials such as ledgers and pens
- information is written down and therefore potentially available for competitors or authorities to see
- cost of keeping records private and secure.

Accurate information is essential and this means that staff who are required to collect information should know its value and why it is being collected. This should be part of the induction and training when new staff learn their job. The entrepreneur should employ people who have the skills and aptitude to do the work, but should also put in place a system of checks to ensure that one person does not have responsibility for a whole area of business activity. For example the person responsible for keeping records of purchases should be different from the person who records use of materials or levels of stocks. The owner or manager should also ensure that all records are kept up to date and where appropriate, the arithmetic is checked for accuracy. There is no single correct way to keep records and individual owners should devise systems that suit their way of working. The examples given below have been found to be successful in small food processing enterprises in Africa and Asia.

2.9.1 Financial and Sales Records

A separate record of the cash that comes into a business and the cash that is used to buy daily items is usefully prepared using a *Cash Book* (Figure 62). Additionally, when entrepreneurs have a bank account, they will require a *Bank Book* to record cheques that have been received and paid, using the same headings as those shown in Figure 62.

Date	Item	Cash in	Cash out	Balance
____	*Description or invoice number*	____	____	____
____	____________	____	____	____
____	____________	____	____	____

Figure 62. - Example of a Cash Book layout

It is important to know how much money the business is owed by debtors at any given time but also how much is owed to creditors. This is particularly important if for example, retailers expect a period of credit before they pay for goods received. The amount of money owed by an enterprise and the amount owing to it can be combined in a single ledger so that a weekly comparison of the difference can be made. Invoices and receipts should be kept together in date order. An example of this type of ledger is an *Accounts Receivable and Payable Book* (Figure 63

Date	Item	Credit given	Balance	Date	Item	Credit taken	Balance
—	*Description or receipt number*			—	*Description or invoice number*		
—	—	—	—	—	—	—	—
—	—	—	—	—	—	—	—
—	—	—	—	—	—	—	—

Figure 63. - Example of an Accounts Receivable and Payable Book

Other books can be used but these are the basic requirement for collecting all financial information needed to prepare monthly profit and loss statements, balance sheets and to check cashflow forecasts. The other information needed to prepare profit and loss statements are records of sales and stock in the storerooms (Figures 64 and 65).

Product Name ________				Batch Number ______	
Date	Customer	Amount Sold	Value	Invoice Date	Payment Date
—	—	*Write in amount in kg or number of packs*	—	—	—
—	—		—	—	—
—	—	—	—	—	—
—	—	—	—	—	—

Figure 64. - Example of a page from a Sales Book

Records that are kept by storekeepers show which products and materials are transferred into and out of the store-rooms. The balance is used to indicate when re-ordering is needed and can also be used to highlight pilferage or other losses that are not accounted for.

Ingredient name ______________________

Date	Amount to store	Amount from store	Process batch number	Balance
___	*Write amounts in kg or number of packs*		___	___
___	___	___	___	___
___	___	___	___	___
___	___	___	___	

Figure 65. - Example of a Storekeeper's Book to keep account of ingredients (similar entries are made for packaging, materials and finished products)

As shown in Figure 66, data from the sales book is totalled to give monthly income. The costs of ingredients, packaging etc., that were used during the month are recorded in the storekeeper's book and other expenses are totalled from the cash book and bank book to calculate the monthly Profit and Loss Account.

The Profit and Loss Account describes how money comes into and leaves a business over a month (or other suitable period of time). This allows the owner to plot the progress of the business and compare the results to those expected in the Business Plan (Section 2.3.4).

However, to obtain a 'snapshot' of the performance of the business at a given moment, a b*alance sheet* is a strong management tool which can help to understand where money came from, how it is used in a business and how it could be better used. An example of a balance sheet from a small wine-making business is shown in Figure 67.

Month: April Item	$	$	
Income from sales		750	*from Sales Book*
Less costs			
Opening stocks	25		
Purchases	55		
Stocks at end of month	30		
= Stocks used during month	50		*from Store-room Ledger*
Gross profit		700	
Less other expenses:			
Salaries	75		
Rent	25		
Supplies	20		*from cash and bank books*
Transport	45		
Marketing costs	25		
Interest repaid to lender	18		
Taxes paid	6		
214			
Net profit		486	

Figure 66. - Example of a monthly Profit and Loss Account

The balance sheet is therefore a statement about the money in a business at a particular time, which shows how the money is being used (the assets) and where it came from (the liabilities). In the above example, the money that remains in the business as unclaimed profits is a main source of working capital. It is important to note that the owner has already taken a salary from the business and that the remaining profit belongs to the business to be used for reinvestment. This picture of the business can be used to determine, for example, whether more stock should be ordered, whether unpaid invoices to retailers should be followed up urgently or whether there are sufficient profits to repay a larger amount from the loan.

Where money came from (Liabilities) ($)		How the money was used (Assets) ($)	
Accounts payable	450	Cash	65
Customer payments	865	Accounts receivable	650
Bank loan	1200	Stocks	600
Owner's capital	500	Equipment	2180
Profits	880	Owner's salary	400
	3895		3895

Figure 67. - Example of a balance sheet for a small wine-maker on 21.3.1997

2.9.2 Production Records

The main reasons for production records are to ensure that quality assurance procedures are in place and operating satisfactorily and to record the use of ingredients and amounts of stock for use in financial accounting. When raw materials are processed, each batch should be recorded in an *Incoming Materials Test Book* (Figure 68). The same layout can be used for recording incoming batches of ingredients and packaging materials, some of which also require inspection on arrival (see Section 2.7.2).

Records should also be kept of the amount and type of raw materials and ingredients that are used and the important processing conditions (e.g. drying times, heating times and temperatures etc.) to ensure that operators mix together the same ingredients in every batch and process them in the same way each time (Figure 69).

Each batch of food should be given a Batch Number which is recorded in stock control books, processing logbooks and product sales records. The batch numbers should be correlated with the product code numbers that are printed on labels or outer cartons. This allows the processor to trace any subsequent faults in a batch of product back to the process or to the raw materials.

Product Name ____________			Batch Number ______	
Raw material*	Supplier	Results of inspection for**: A	B	C
Write in either 'Pass/Fail or observations on quality				
______	______	______	______	______
______	______	______	______	______
______	______	______	______	______

Figure 68. - Example of an Incoming Materials Test Book

* This information should also be recorded for other ingredients and packaging materials

** Different tests are written in place of ABC according to the types of materials being inspected

Product ____________			Batch Number ______
Ingredients:	Target	Check	Changes from target
Pulp	*Write in amounts, times etc.*		______
Sugar	______	______	______
Citric Acid	______	______	______
Pectin	______	______	______
Batch weight	______	______	______
Boiling time	______	______	______
Solids content of product			

Figure 69. - Example of a Process Logbook for jam production

In summary, therefore, record keeping is a management tool to help the owner to know the state of a small fruit and vegetable enterprise at any time and to have reliable information on which to base his or her plans for development of the business.

Appendix I

Basic Rules for Hygiene, Sanitation and Safety

Facilities Required in the Processing Room

- A changing room and separate area for clothing and shoes that are not worn for work, which may be combined.
- Separate hand-washing facilities in a convenient place for staff to use, with soap, clean water, nail brushes and clean towels.
- Toilets, which should be separated from the processing room by two doors or located in a nearby building.
- First aid materials in an accessible place.
- Protective clothing, including gloves and shoes if appropriate, should be provided and laundered regularly.
- Cleaning chemicals should be stored away from the processing room.

Ways of working

- Clean all areas, including toilets, washing facilities and storerooms every day.
- Use the correct chemicals for cleaning equipment, make sure there are no food residues and rinse the equipment with clean water.
- Make sure all cleaning cloths are washed and boiled each day. Do not hang them on equipment, product or window ledges to dry.
- Clean equipment 'as you go' and do not leave dirty equipment until the end of the day before cleaning it.
- Keep the area around the processing room clean and tidy, keep grass cut short.
- Do not allow wastes to accumulate but remove them as they are produced. Clean up any spillages as they occur. Wastes should be collected in special bags or bins that are not used for anything else and taken to be dumped or burned away from the processing site.
- All animals should be prevented from entering the site and especially from entering the processing area or storerooms.
- Visitors should not enter the processing room without protective clothing and under supervision.
- Wear protective clothing which does not have loose ends that could get caught in machinery. Make sure that it is kept clean.
- Wear a hat that completely covers the hair, do not comb your hair in a processing room or storeroom.
- Do not wear wristwatches or jewellery as these are a source of infection and could get caught in machinery.
- Cover all cuts, burns, sores and abrasions with a clean, waterproof dressing.

- Do not smoke, use chewing gum, snuff or betel nuts in any room where there is open food because bacteria can be transferred from the mouth to the food.
- Do not spit in any part of a processing room or storeroom.
- Wash hands and wrists thoroughly with soap after
    - entering the processing room,
    - using the toilet,
    - between handling raw and cooked food
    - eating, smoking, coughing, blowing your nose or combing your hair
    - handling waste food, rubbish or cleaning chemicals.
- Dry them on a clean towel before handling food again.
- Do not use hand-washing facilities for food preparation.
- Keep finger nails cut short.
- Do not wear perfume or nail varnish as these can contaminate products.
- Do not lick fingers (e.g. when picking up sheets of wrapping paper).
- Do not handle any food if you have sores, boils, septic spots, a bad cold, sore throat or a stomach upset. Report any of these to the manager and do alternative work.
- Do not cough or sneeze over food.
- Keep food covered wherever possible.
- Keep all food, tools and equipment off the floor.
- Keep ingredients in sealed containers.
- Do not use unsuitable, broken or dirty equipment.
- Report any signs of insects, rodents or birds to the manager.

Safe working

- Do not leave metal stirrers in boiling liquids.
- Do not leave handles of boiling pans over the heat source.
- Carry knives with the point down, do not try to catch falling knives.
- Cut fruits and vegetables on a board and not using the other hand.
- Do not use a damp cloth to carry hot pans.
- Wear shoes that protect your feet from falling or hot objects.
- Cover burning cooking oil with a damp cloth, never use water to put out flames.
- Do not use gas burners in direct sunlight as the flames can become invisible.
- Do not carry large containers of hot food on your own, get assistance.
- Do not put cleaning fluids into old food containers.

(Adapted from: *The Food Hygiene Handbook*, by R.A Sprenger)

Appendix II

Checklist of Entrepreneurial Characteristics

Ask yourself the questions below to see whether you are the sort of person that could set up and operate a food business.

		YES	NO
1.	I think that a routine pattern of life with regular working hours suits me best	.....	.....
2.	I have always thought and acted by myself	.....	.....
3.	Some days I seem to achieve nothing	.....	.....
4.	It is not good to start something unless you are going to finish it	.....	.....
5.	I am much happier when I do not have to rely on other people	.....	.....
6.	I often feel that I am the victim of events that I can not control	.....	.....
7.	In any bad situation I always get something good from it	.....	.....
8.	It is very important to me that people recognise my success	.....	.....
9.	I am not too ambitious so that I can avoid being disappointed	.....	.....
10.	I am prepared to take risks only after I have thought about all of the possible consequences	.....	.....
11.	When I talk to a senior person I do not usually say what I mean	.....	.....
12.	People often tell me that I am good at understanding their point of view	.....	.....
13.	The amount of money that I earn is more important than how hard I work to earn it	.....	.....
14.	I usually work later than I planned	.....	.....

Answers:

Yes to questions 2,4,7,8,10,12,14
No to questions 1,3,5,6,9,11,13
Score 5 points for every correct answer and 0 points for every incorrect answer. The higher the score the more likely you will be a successful entrepreneur, but a score above 35 is a good sign.
(From: *Starting a Small Food Processing Enterprise*, by Fellows, Franco and Rios, IT Publications)

Appendix III

Feasibility Study Checklist

1 **Name of Business**
Location (accessibility to raw materials, labour and markets)
Description of products and benefits

2 **Market Feasibility**
Target market
Volume and value of unsatisfied demand
Present supply and competition
How will the demand be met?
Price, packaging, distribution, promotion,
Projected sales, terms of business, organisation

3 **Technical Feasibility**
The manufacturing process
Raw materials required and their sources
Labour and skills required
Equipment and utilities required
Production schedule (for one year plus optimum production volume)
Production costs

4 **Financial Feasibility**
Business costs
Sales income
Cash flow projection
Profit and loss projection
Financing needs and sources

5 **Major Assumptions**
Legislation
Government policy (subsidies, incentives and taxation)

6 **Risks and Benefits**

7 **Will the Business be Successful?**
an overall assessment of feasibility

Appendix IV

Institutions that Support Small Scale Food Processing

Africa

Bostwana. Botswana Technology Centre (BTC), PO Bag 0082, Gaborone.

Ethiopia. Food Research and Development Centre, Ethiopian Food Corporation, PO Box 5688, Addis Ababa.

Gabon. Institute de Recherche Technologique, BP 14070, Libreville-Akebe.

Ghana. Technology Consultancy Centre (TCC), University of Science and Technology, Kumasi.

Ghana. Ghana Regional Appropriate Technology Industrial Service (GRATIS), PO Box 151, Tema.

Kenya. Federation of Kenyan Employers (FKE), PO Box 48311, Nairobi.

Kenya. Approtech, PO Box 10973, Nairobi.

Kenya. Technoserve, PO Box 14821, Nairobi.

Malawi. TCC/PHN Women in Development Project, PMB 3, Namadzi.

Malawi. Garry Whitby Associates, PO Box 1836, Lilongwe.

Malawi. Malawi Enterprise Development Institute (MEDI), PMB 2, Mpnela.

Malawi. Small Enterprise Development Organisation of Malawi (SEDOM), PO Box 525, Blantyre.

Namibia. Development Centre for Research Information Action in Africa (CRIAA), 22 Johan Albrecht Street, Windhoek.

Nigeria. Dept. Food Science and Technology, Federal University of Technology, PMB 1526, Owerri.

Nigeria. Dept. Food Technology, Univeristy of Agriculture, PMB 2373, Makurdi.

Nigeria. International Institute of Tropical Agriculture (IITA), PMB 5320, Ibaban.

Republic of South Africa. Council for Scientific and Industrial Research (CSIR), PO Box 395, Pretoria 0001.

Sudan. Food Processing Research Centre, Khartoum.

Tanzania. Small Industries Development Organisation (SIDO), PO Box 2476, Dar-es-Salaam.

Tanzania. National Food Control Commission, PO Box 7601, Dar-es-Salaam.

Tanzania. Dept. Food Science, Sokoine University of Agriculture, Morogoro.

The Gambia. Gambia Food and Nutrition Association (GAFNA), PO Box 111, Banjul.

Uganda. Uganda Manufacturers Association, PO Box 6966, Kampala.

Uganda Uganda Small Scale Industries Association (USSIA), PO Box 2344, Mbale.

Uganda. National Agricultural Research Organisation (NARO), PO Box ??, Kampala.

Uganda. Dept. Food Sciences and Technology, Makerere University, PO Box 7062, Kampala.

Uganda. Nile Vocational Institute, PO Box 1829, Jinja.

Zimbabwe. Ranche House College, PO Box 1880, Harare.

Asia

Bangladesh. Bangladesh Rural Advancement Committee (BRAC), 66 Mohakhali Commercial Area, Dhaka 12,

India. Academy of Development Science (ADS) Kashele, Karjat Taluka, Raigad District, Maharashtra 410 201,

India. All India Food Preservers Association (AIFPA), 206 Aurobindo Place, Hauz Khas, New Delhi 110016,.

India. Small Industries Research Institute (SIRI), PO Box 2106, New Delhi 110007,

India. Action for Food Production (AFPRO), C52, ND South Extension II, New Delhi 16,

India. Central Food Technology Research Institute (CFTRI), Mysore 570 013, Maharashtra State,

India. Dept. Food Science and Technology, JN Agricultural University, Jabalpur, Madhya Pradesh 482004,

India. Agriculturl and Processed Foods Export Development Authority (APEDA), 6 Bhikaji Cama Place, New Delhi 66,

India. Society for Development of Appropriate Technology (SOTEC), 182 Civil Lines, Bareilly, Uttar Pradesh,

Indonesia. Bandung Institute of Technology, PO Box 276, Bandung,

Nepal. Centre for Rural Technology, PO Box 3628, Kathmandu,

Sri Lanka. National Packaging Centre, Lower Chatham Street, Colombo 1,

Sri Lanka. Mahaweli Enterprise Development Programme, Unity Plaza, Colombo 4,

Sri Lanka Sri Lanka Standards Institute (SLSI), 53 Dharmapala Mawatha, Colombo 3,.

Sri Lanka. Industrial Development Board (IDB), 615 Galle Road, Moratuwa,

Sri Lanka. Council for Scientific and Industrial Research (CISIR), PO Box 787, Colombo 7,

Thailand. Food Technology Laboratory and Institute of Food Research and Product Development, Kasatsaart University, both at 196 Phahonyothin Road, Bangkok 10900,

West Indies and South Pacific

Antigua, Chemistry and Food Technology Division, Ministry of Agriculture, Fisheries and Lands, Dunbars, West Indies.

Papua New Guinea. Appropriate Technology Development Institute, Papua New Guinea University of Technology, PO Box 793, Lae,

Europe

Austria. United Nations Industrial Development Organisation (UNIDO), PO Box 300, A-1400 Vienna.

Denmark. Danish International Development Assistance (DANIDA), 2 Asiatisk Plads, DK-1448, Copenhagen.

France. Groupe de Recherche et d'Echanges Technologiques (GRET), and Technologie et Partenariat en Agroalimentaire (TPA), 213 rue Lafayette, Paris 75010.

Finland. Finnish International Development Agency (FINNIDA), Katajanokanlaituri 3, FIN-00160, Helsinki.

Germany. Fordergesellschaft fur angepasste Techniken in der Dritten Welt mbH (FAKT), Stephan-Blattmannstrasse 11, D-78120, Furtwangen.

Germany. German Appropriate Technology Exchange (GATE/GTZ), Postfach 5180, D-6236 Eschborn 1.

Italy. United Nations Food and Agricultural Organisation (FAO), Via delle Terme di Caracalla, 00100, Rome.

Switzerland. International Labour Office (ILO), Employment and Technology Branch, CH-1211 Geneva.

Switzerland. Swiss Centre for Development Cooperation in Technology and Management (SKAT), Vadianstrasse 42, CH-9000, St. Gallen.

The Netherlands. Technology Transfer for Development (TOOL), Sarphatistraat 650, 1018 AV Amsterdam.

The Netherlands. International Agricultural Centre (IAC), 11 Lawickse Allee, 6701 AN Wageningen.

The Netherlands. KIT Royal Tropical Institute, Mauritskade 63, 1092 AD Amsterdam.

The Netherlands. Agromisa, PO Box 41, 6700 AA Wageningen.

The Netherlands. Technical Centre for Agriculture and Rural Cooperation ACP-EU (CTA), Postbus 380, 6700 AJ, Wageningen.

UK Midway Technology Ltd., St Oswalds Barn, Clifford, Hay on Wye, Herefordshire.

UK. Intermediate Technology Development Group (ITDG), Myson House, Railway Terrace, Rugby, CV21 3HT, (ITDG also has offices in Bangladesh, Sri Lanka, Sudan, Zimbabwe, Kenya and Peru).

UK. Natural Resources Institute (NRI), Central Avenue, Chatham, Kent, ME4 4TB.

Latin America

Cost Rica. Centro de Investigaciones en Tecnologias de Alimentos, Universidad de Costa Rica, CP 2060.

Guatemala. Centro de Estudios Mesoamericanos sobre Technologias Apropriadas (CEMAT), Apartado Postal 1160, 28 Avienda 18-80, Zona 10-01010, Ciudad de Guatemala.

Guatemala. Institute de Nutrition de Centro America y Panama (INCAP), Calzada Roosevelt Zona II, Partado Postal 1188, Ciudad de Guatemala.

North America

Canada. International Development Research Centre (IDRC), Box 8500, Ottawa, K1G 3H9.

USA. Appropriate Technology International (ATI), 1828 L Street NW., Washington DC 20036.

USA. Post-Harvest Institute for Perishables Information Centre, University of Idaho Library, Moscow, Idaho 83843.

USA. United Nations Development Fund for Women (UNIFEM), UN Plaza, New York, NY 10017.

USA. Technoserve, 49 Day Street, Norwalk, Conneticut 06854.

USA.. Compatible Technology Inc., 5835 Lyndale Avenue S., Minneapolis MN 55419.

NB: UN organisations can be contacted via United Nations Development Programme (UNDP) offices in most countries and bilateral donors can be contacted via Embassies or High Commissions of the countries concerned.

Bibliography

Fruit and Vegetable Processing and Products

Conservation des Fruits a Petite Echelle, 1990, Bureau International du Travail, Geneva, Switzerland. (ISBN 92 2 206403 8)

Domestic Preservation of Fruit and Vegetables, 1954, Ministry of Agriculture and Fisheries, Her Majesty's Stationery Office, 49 High Holborn, London, UK.

Food Processing Technology, Fellows, P.J., 1993, Woodhead Publishing, Cambridge, UK

Fruit and Vegetables, MacDonald,I. and Low,J.,1984, IT Publications, London, UK., (ISBN 0 237507 900)

Fruit and Vegetable Processing Appropriate Technology in Costa Rica; A Case Study, Quiros, R.D., Madrigal,A., Samuals, L., Aguilar, A., Orfiż, F., Fernandez, R. and Cooke, R., 1980, Tropical Science, 22 (2).

Fruit Processing, Axtell,B., Kocken E., and Sandhu, R., 1993, Food Cycle Technology Sourcebook, IT Publications, London, UK

Home Preservation of Fruit and Vegetables, Ministry of Agriculture , Fisheries and Food, 1969, Bulletin 21, Her Majesty's Stationery Office, 49 High Holborn, London WC1., UK.

Home Scale Processing and Preservation of Fruits and Vegetables, 1977, Central Food Technology Research Institute, Mysore, India.

Integrated Food Science and Technology for the Tropics, 1985, Ihekoronye, A.I., and Ngoddy, P.O., Macmillan Press Ltd., London, UK.

Le Point sur la Transformation des Fruits Tropicaux, Bidault, B. and Gattegno, I., 1984, GRET (Groupe de Recherche et D'echanges Technologiques), Paris, France.

Processing Tropical Crops, Asiedu, J.J., 1989, MacMillan Press Ltd, London, (ISBN 0-333-44857-X)

Rural Home Economic Food Preparation, Series 1, Food Preservation, Series 2, Labour Saving Ideas, Series 3, FAO, Via delle Terme di Caracalla, 00100 Rome, Italy.

Small Scale Food Processing: a guide to appropriate equipment, P. Fellows and A Hampton, 1992, IT Publications, 103-105 Southampton Row, London WC1B 4HH, UK (ISBN 1-85339-108-5)

The Orange Hill Estate: a Successful Small Industry in St. Vincent, Axtell, B., 1983, AT Journal, 10, (2).

Tomato and Fruit Processing: an example of a village factory, de Klein, G., 1993, TOOL Publications, Amsterdam, Holland. (ISBN 90 70857 31 6)

Traditional and Non-Traditional Foods, Ferrando,R., 1981, FAO Publications, Via delle Terme di Caracalla, Rome, Italy. (ISBN 92-5-100167-7)

Traditional Food Technologies, Fellows, P., (Editor), expected 1997 from IT Publications, London, UK.

Tropical Fruit Processing, Jagtiani,J., Chan, H.T. and Sakai, W.S., 1988, Academic Press, San Diego, California, USA. (ISBN 0 12 379990 2)

Vegetable Production in the Tropics, Williams, C.N., Uzo, J.O. and Peregrine, W.T.H., 1991,
Longman Press, London, UK. (ISBN 0 582 60609 8)

Vegetables in the Tropics, Tindall,H.D., 1983, MacMillan Press Ltd., London (ISBN 0 333 24266 1)

Drying

A solar food drier for Bangladesh, Clark, C.S., 1982, Appropriate Technology,8, 4, 14-16

Dates: Handling, Processing and Packing, Dowson, V.H.W. and Aten, A., 1962, FAO, Rome, Italy.

Date Palm Products, Barreveld, W.H., 1993, Agricultural Services Bulletin No. 101, FAO, Rome, Italy.

Dehydrated Fruit Report, McBean, M.G., Joslyn, M.A. and Nury, F.S., 1970, Commonwealth Scientific Industrial and Research Organisation, Sydney, Australia.

Development of a forced convection solar drier for red peppers, Trim, D.S. and Ko, H.Y., 1982, Tropical Agriculture (Trinidad) 49,4,319-323

Drying crops with solar heated air, Buelow, F.H., 1961, Proceedings of UN conference on New Sources of Energy, FAO, Rome, Italy.

Effect of drying on the nutritive value of foods in Kenya, Gomez, M.I., 1982, in Food Drying; IDRC-195e (Ed.) Yaciuk G. International Development Research Centre, Ottawa, Canada.

Osmotic dehydration - a cheap and simple method of preserving mangoes, bananas and plantains, Hope, G.W. and Vitale, D.G., 1972, Report IDRC 004 el. International Development Research Centre, Ottawa, Canada.

Osmotic dehydration of fruits, Ponting, J.D., Watters, G.G., Forrey,R.R. and Stanley, W.L., 1966, Journal of Food Technology, 1,(10),125-128.

Pretreatment for Solar and Hot-air Drying, Wagner, C.J., Coleman, R.L. and Berry, R.E., 1980, Proceedings of Florida Horticultural Society, 93,336-338.

Small scale solar crop dryers for tropical village use - theory and practical experience, Grainger.W., Othieno, H. and Twiddle, J.W., 1981, Proceedings of ISES solar world forum, Brighton, UK.

Solar Dryers - Their Role in Post-Harvest Processing, 1987, Commonwealth Science Council, London, UK.

Solar drying of crops and foods in humid tropical climates, McDowell, J., 1973, Report CFNI-T-7-73, Caribbean Food and Nutrition Institute, Kingston, Jamaica.

Solar Drying: Practical Methods of Food Preservation,1986, ILO, Geneva, Switzerland (ISBN 92 2 105357 1)

Solar food drying: A rural industry, Clark, C.S., 1981, Renewable Energy Review Journal, 3,1,23-26

The blanching of vegetables and fruit, Voirol, F., 1972, Food Process Industries, August.

The practicality of solar drying of tropical fruits and marine products as income generation for rural development, Pablo, I.S., 1978, Proceedings of UNESCO solar drying workshop, Manila, Philippines.

Try Drying IT, Case Studies in the Dissemination of Tray Drying Technology (including Bangladesh & Guatemala), Axtell, B.L., and Bush, A., 1991, IT Publications, London, UK. (ISBN 1 85339 039 9).

Use of solar energy for banana drying, Bowrey, I., Buckle, R.G., Hamey, K.A. and Payenayotin, P., 1980, Food technology in Australia, 32, (6), 290-291.

Preserves

Jam Manufacture, Rauch, G.H., 1965, Leonard Hills Books, London, UK.

Jams, Pickles and Chutneys, Mabey, D.& R.,1985, Penguin Publications, Harmondsworth, UK.

Juices and Fermented Products

Fruit and Vegetable Juice Processing, Nelson, P.E. and Tressler, D.T., 1982, AVI Publications, Conn., USA.

Fruit Juice Processing, Bielig, H.J., 1973, FAO Agricultural Services Bulletin 13, FAO Publications, Via delle Terme di Caracalla, 00100 Rome, Italy.

Home Wine Making and Brewing, Turner, B.C., 1970, Boots Company Ltd., London, UK.

Kick start for Village Vinegar in Papua New Guinea, Adams, M.R., 1982, AT Journal, 9, (2), IT Publications, London, UK.

Microbiology of Fermented Foods, Wood, B.J.B,(ed), 1985, Elsevier, London, (ISBN 0 85334 332 2).

Small Scale Production of Vinegar from Bananas, Adams, M.R., 1980, TDRI Post Harvest Technology Publication, G132, Natural Resources Institute, Chatham, Kent, UK.

Pickles and Sauces

Preservation of Vegetables by Salting or Brining, US Department of Agriculture Farmers Bulletin, No.1932., USDA.

Pickle and Sauce Making, Binstead R., Devey, J.D. and Dakin, J.C., 1971, Food Trade Press Ltd., London, UK.

Feasibility Studies and Business Development

Consultancy for Small Businesses: the concept, training the consultants, Harper, M., 1976, IT Publications, London, UK (ISBN 0-903031-42-6)

Do Your Own Scheme, a Manual for the Entrepreneur, Small Business Promotion Project, Nepal Ministry of Industry and GTZ)

Doing a Feasibility Study: training activities for starting or reviewing a small business, Suzanne Kindervater (Editor), 1987, OEF International, 1815 H Street NW, 11th Floor, Washington, DC 20006, USA. (ISBN 0-912917-07-5).

Entrepreneurship for the Poor, M Harper, 1984, IT Publications, London, UK.

Improving Small-Scale Food Industries in Developing Countries, Edwardson, W. and MacCormac, C.W., (Editors), 1986, IDRC Publications, PO Box 8500, Ottawa, Canada K1G 3H9. (ISBN 0-88936-398-6)

Monitoring and Evaluating Small Business Projects: a step by step guide for private development organisations, Buzzard, S. and Edgcomb, E., (Editors), 1992, PACT, 777 United Nations Plaza, New York, NY 10017, USA (ISBN 0-942127-00-5).

Small Business in the Third World, M Harper, 1984, J Wiley & Sons Ltd, Chichester, UK, (ISBN 0-471-90474-0).

Small Business in the Third World: Guidelines for Practical Assistance, Harper, M., IT Publications, London, UK. (ISBN 0 471 90474 0).

Starting a Small Food Processing Enterprise, Fellows, P., Franco,E. and Rios, W., 1996, IT Publications, London, UK. (ISBN 1 85339 323 1).

Women and Small Business, International Women's Tribune Centre, 1984, Tribune Centre, 777 UN Plaza, New York, NY 10017, USA.

Management and Finance

Basic Accounting for Small Groups, Cammack, J., 1992, Oxfam, 274 Banbury Rd, Oxford, UK. (ISBN 0 85598 148 2).

Consultancy for Small Businesses, Harper, M., 1977, IT Publications, London, UK. (ISBN 0 903031 42 6).

Doing a Feasibility Study: Training Activities for Starting or Reviewing a Small Business, Kindervatter, S., 1991, UNIFEM, New York, NY., USA.

Entrepreneurs Handbook, 1981, Technonet Asia, via IT Publications, London, UK.

Food and Drink - Good Manufacturing Practice: a guide to its responsible management, 1991, Published by IFST, 5 Cambridge Court, 210 Shepherd's Bush Road, London W6 7NL, UK (ISBN 0 905367 08 1)

How to Read a Balance Sheet, Halsall, J.J.H., 1986, ILO, Geneva, Switzerland.

Improve Your Business, Dickson, D.E.N., 1986, International Labour Office, Geneva, Switzerland.

Management for Commercial Analysis of Small Scale Projects, Jackelen, H.R.,1983, Appropriate Technology International, 1724 Massachusetts Avenue, N.W. Washington, DC20036, USA.

Manual of Practical Management for Third World Rural Development Associations, Vol.2., Financial Management, 1989, IRED, Geneva, Switzerland.

Where Credit is Due: Income Generating Programmes in Developing Countries, Remenyi, J., 1991, IT Publications, London, UK. (ISBN 1 85339 079 8).

Marketing

Marketing Strategies: Training Activities for Third World Women Entrepreneurs, OEF International, via IT Publications, London, UK. (ISBN 0 912917 08 3).

Marketing Strategy: training activities for entrepreneurs, Kindervater, S. and Range, M., 1986, OEF International, 1815 H Street NW, 11th Floor, Washington, DC 20006, USA. (ISBN 0-912917-08-3).

Opening the Marketplace to Small Enterprise, de Wilde, T., Schreurs, S. and Richmond, A., 1991, IT Publications, London, UK. (ISBN 1 85339 093 3)

Success with Marketing Diversified Enterprises, Ministry of Agriculture, Fisheries and Food, 1996, MAFF Publications, London, SE99 7TP, UK

Quality Assurance and HACCP

A practical approach to quality control, Caplen, R.H., 1982, Hutchinson Publishing Group, London, UK.

Export Quality, World directory of Standardisation and Quality Assurance Related Institutions, Anon, 1992, ITC/UNCTAD/GATT, Geneva, Switzerland.

Fundamentals of quality control for the food industry, Kramer, A. and Twigg, B.A., 1962, AVI Publishing Co., Westport, Conn. USA.

HACCP - A Practical Approach, Mortimer, S. and Wallace, C., Chapman and Hall, London, UK.

HACCP in street vending in developing countries, Bryan, F.I., 1993, Food Australia, 45. (2).

Hazard Analysis Critical Control Point Evaluations, Bryan, F.l., 1992, WHO, Geneva, Switzerland.

How to HACCP, an illustrated guide, Dillon, M and Griffith, C.,1995, MD Associates, Unit 43, Jackson Place, Wilton Rd, Industrial Estate, Grimsby, DN36 4AS, UK.

Manuals of Food Quality Control, Vols 1-9, FAO Publications, Via delle Terme di Caracalla, 00100 Rome, Italy. (Vol.3: *Commodities*, Martin,P., 1979, Vol 4: *Microbiological Analysis*, Refai, M.K., 1979, Vol 5: *Food Inspection*, 1984, Vol 6: *Food for Export*, Dhamija, O.P., 1979, Vol 7 *Food Analysis*, 1986, - *general techniques, additives, contaminants and composition,* Vol 8: *Food Analysis*,1986, - *quality, adulteration and tests for identity*, Vol 9 *Introduction to Food Sampling*, 1988)

Procedures to implement HACCP systems, 1991, IAMFES, Published by International Association of Milk, Food and Environmental Sanitarians, USA.

Quality Control for the Food Industry, 1991, ITC UNCTAD/GATT, Geneva, Switzerland.

Quality Control in Fruit and Vegetable Processing, Board, P.W., 1988, FAO Food and Nutrition Paper No. 39, FAO Publications, Via delle Terme di Caracalla, 00100 Rome, Italy. (ISBN 92-5-102421-9)

Safe Processing of Foods, Shapton, D.A. and Shapton, N.F., 1993, Butterworth-Heinemann, Oxford, UK.

Sanitation and Hygiene

Affordable Water Supply and Sanitation, Pickford, J., Barker, P., Coad, A., Dijkstra, T., Elson, B., Ince,M., and Shaw, R (Editors), 1995, IT Publications, London, UK. (ISBN 1 85339 294 4).

Codes of hygienic practice of the Codex Alimentarius Commission, Anon, Joint FAO/WHO Food Standards Programme, FAO Publications, Via delle Terme di Caracalla, 00100 Rome, Italy.:
1 *Canned fruit and vegetable products* (CAC/RCP 2, 1969)
2 *Code of hygienic practice for dried fruits* (CAC/RCP 3, 1969)
3 *General principles of food hygiene* (CAC/RCP 1, 1979).

Disinfection of Rural and Small-Community Water Supplies, Anon, 1989, Water Research Centre, Medmenham, Bucks, UK.

Food Poisoning and Food Hygiene, Hobbs, B. and Roberts, D., 1987, Edward Arnold Ltd, 41 Bedford Square, London WC1B 3DQ, UK (ISBN 0-7131-4516-1).

Hygienic design and Operation of Food Plant, Jowitt, R., (Editor), 1980, Ellis Horwood Ltd, Cooper Street, Chichester, PO19 1EB, UK (ISBN 0-85312-153-2) (NB for larger scale food processing)

Making Safe Food (Book and Posters), Fellows, P. and Hidellage, V., 1992, IT Publications, London, UK.

Safe Drinking Water, Howard, J., 1979, Oxfam Technical Guide, Oxfam, Oxford, UK.

The Food Hygiene Handbook, Sprenger, R.A., 1996, Highfield Publications, Doncaster, DN5 7LY, UK (ISBN 1 871912 75X).

Packaging Materials and Labelling

Appropriate Food Packaging, Fellows, P. and Axtell, B., 1993, TOOL Publications, Sarphatistraat 650, 1018 AV Amsterdam, The Netherlands. (ISBN 90 70857 28 6).

Packaging, Obi-Boatang, P. and Axtell, B.L., 1995, Food Cycle Technology Sourcebook: IT Publications, London, UK.

Basic Science Food Composition and Chemical Analysis

Food: the Chemistry of its Components, Coultate, T.P., 1984, The Royal Society of Chemistry, Burlington House, London W1V 0BN, UK. (ISBN 0-85186-483-X).

Modern Food Microbiology, Jay, J.M., 1978, D van Nostrand, New York, (ISBN 0 442 24127 5).

Pearson's Chemical Analysis of Foods, Egan, H., Kirk, R.S., and Sawyer, R., 1981, Churchill Livingstone, London. (ISBN 0 443 02149 X).

The Composition of Foods, Paul, A.A. and Southgate, D.A.T., 1985, Her Majesty's Stationery Office, 49 High Holborn, London, UK., (ISBN 0 11 450036 3).

Food Security and Community Development

Approaches to Participation in Rural Development, Oakley, P. and Marsden, D., 1984, ILO, Geneva, Switzerland.

Building Sustainable Communities: Tools and Concepts for Self-reliant Economic Change, Morehouse, W., 1989, Bootstrap Press via IT Publications, London, UK. (ISBN 0 942850 11).

Communication Skills for Rural Development, MacDonald, I. and Hearle, D., 1984, Evans Publishing via IT Publications, London, UK.

Developing Technologies for the Rural Poor, Biggs, S. and Grosvenor-Alsop, R., 1984, IT Publications, London, UK.

How to Run a Small Development Project, Geneva Group, 1986, IT Publications, London, UK. (ISBN 0 946688 47 8).

Makala Ya Mafunzo: Skill Development for Self Reliance, ILO/SDSR, PO Box 60598, Nairobi, Kenya.

Providing Food Security for All, Alamgir, M. and Arora, P., 1991, IT Publications, London UK. (ISBN 1 85339 115 8).

Training Village Entrepreneurs: Guidelines for Development Workers, 1986, Skills for Progress, via IT Publications, London, UK.

Village Technology Handbook, VITA, 1970, VITA College Campus, Schenectady, New York 12308, USA.

Note: *FoodChain* is a publication concerned with small scale food processing that is free to subscribers in developing countries, published by Intermediate Technology Development Group, Railway Terrace, Rugby, CV21 3HT, UK and has regular features on many of the above topics.

Index

A

Acetic acid -see vinegar.
Acetic acid bacteria, **56**, 60
Acidity, 27, 33, 38-40, 48-49, 56, 133, **139**
Activity chart, 82-83
Additives, 74-75, 97
Agreements with retailers, 16, **121-123**
Air lock, 57
Alcohol, 3, 12, 56-61, 97, 114
Apples, 43
Apricots, 8,10
Ascorbic acid, 7, 8, 10
Associations of producers, 10-11, **16-17**
Average weight system, 98

B

Banana, 8, 12, 25-26, 56-58, 138
Balance sheet, 159
Baobab, 12
Beans, green, 10, 12
Beer, fruit, 7, 12, 57
Benefits of food processing, 4, 6
Bibliography, 173-180
Blanching, 3, 29, **30-33**, 110, 113
Boiling pans, stainless steel, 30, 39, 42-46, **47**, 51-54, 78, 113
Bottles - see glass,
Bottling of fruit, 3, 24, **27-29**, 48
Brands, brand image, 71-72, 152-154
Breakeven point, 88-89
Brix, 60, 139
Buildings, 102-109
    ceilings, 102
    floors, 103
    layout of equipment, 103, 106-109
    lighting and power, 104
    roofs, 102
    walls, 102
    water supplies, 104
    windows and doors, 102
Business plan, **91-92**, 158

C

Cabbage, 8, 10, 13-14, 42
Cabinet - dryer, 112
Canning, 3, 24, **27-29**
Candied fruits - see crystallised fruits
Capital costs, 63
Carbonated drinks, 47
Carotene, 7
Carrot, 8,10
Case hardening, 112
Cashflow, 15, 20, **85-90,** 120-121, 127, 157
Checkweighing, 142
Chillies, 14, 39, 45, 65
Chilling, 3
Citrus fruits, 43, 48, 51-52, 114
Citrus reamers, 114
Coconut, 32, 35
Competitors, 20, 23, 63, **70-75, 87, 91,** 149, 152-154, 156
Composition of food, legislation, 95-96
Concentration by boiling, 3, 12, 24, **42-46, 54**, 78
Contracts with retailers, 16, **121-123**
Contracts with suppliers, 79, **119-121**
Cordials, 24, **50-51**, 96
Critical control points, 133-135
Crystallised fruits, 24, 29-30, **33-34**
Cucumber, 42
Customers, 19, 65-68, 71-74, 92, 95, 97-98

D

Demand for processed food, 20, 63
Distillation, 24, **56, 61,** 114-115

Distribution of products, 72-75, 121-123, **141-143**, 149
Dried fruits and vegetables, 2, 7, 24, **29-37**, **110-112**
Dryers, 3, 31, **34-35, 111-112**
Drying, 3, 11, 24-25, **29-37**, 108, **111-112**

E

Enterprise development, 5
Entrepreneurial characteristics, 163
Equilibrium relative humidity, 35, 140
Equipment for processing, 25, 28, 30-31, 39, 42-43, 46, 48, 51, 53, 59-60, **80-81**, 84, **107-116**, 127
Equipment for analyses, 137-145

F

Faults in jam making, 55
Feasibility studies, 63-92
- checklist, 165
- financial, 85-90
- markets, 64-84
- technical, 76-84

Fermentation, 3, 12-13, 24, 38-42, 56-60, 114
Filling, 25, 28, 31, 35, 39, 42, 46, 48, **50**, 51-54, 59-60, **115-117**, **142-143**
Finance planning, 90
Fixed costs, 86-87
Food
- development organisations, 167-171
- laws, 93-99
- processors, electrical, 30
- safety, 130-145
- security, 4-5, **8-11**

Freezing, 3
Fruit , cheese, 24, **52-54**
Frying, 24-26
Fuels, 111-112, **115**

G

Garlic, 39, 52, 96
Gas, fuel, 115
Ginger, 52
Glass, jars and bottles, 27-29, **37**, 39-42, 46-49, 51, 53, 59-60, 117, **142-144**
Go/no-go rings, 142-144
Grapefruit, 52
Grapes, 10, 56

H

HACCP, 130-145
Harvesting, 25, 28, 30, 39, 42-43, 46-48
Headspace gauge, 56, 142
Heat sealers, 25, 31, **116**
Home processing11, 17
Humidity, 35, 111-112, 118, 140
Hydrometers
- alcohol, 59, 114, 138-140
- brine, 42, 139-140

Hygiene, 97-99, **104-106**, 144-145, **161-162**

I

Income, 87-88
Institutions, support for food processing, 167-171

J

Jakfruit, 21, 25
Jam, 24, **52-55**, 66, 96
- faults, 55

Jars - see glass
Jellies, 24, **52-54**, 96
Juice, 24, **47-50**, 54, 58, 96, 113

K

Ketchup, 96

L

Labelling, 28, 42, 46-48, 51, 59-60, **97-98, 152-154**
Leather, fruit, 7, 24, 30
Legislation, 93-99
- business registration, 92-94
- composition, 52, 95-96
- food laws, 95-99
- hygiene and sanitation, 97-98
- labelling, 97, 152-154
- weights and measures, 98
Lemons, 43, 52
Limes, 14, 42-43, 52
Limited liability, 93
Liquid fillers, 30, 39, 42, 46, 48, 51, 53, 59-60
Losses, during processing, 14, **77-79**, 81

M

Maintenance of equipment, 127
Management
- financial, 85-90
- equipment maintenance, 127
- health and safety, 129-130
- product quality, 134-145
- product safety, 130-133
- production, 125-130
- quality assurance, 130-145
- staff, 128-129, 134
- stock control, 143
Mangoes, 8, 14, 38-42, 77-78, 84
Markets, **64-72**, 84, 92
Market
- analysis, **64-72**, 147-150
- mix, 151
- niches, 148
- research, 64-72
- share, 70-73, 84
- segments, 147-150
- size and value, 67-72, 84
- strategies, 151-154
- surveys, 64-67
Marketing, 147-154
Marmalade, 24, 36, **52-55**, 96, 138
Melons, 48, 56
Moisture36, **140-141**

N

Nectars, 24, 96,
Net weight, 96, 142-143, 153
Nutrition, 5-10

O

Okra, 8
Onions, 7, 14, 40, 80, 89, 96
Operating costs, 86-87
Oranges, 8, 43, 52
Osma-sol drying, 29
Owner's equity, 85

P

Packaging, 25, 28, 31, **35-37**, 39, 42, 46, 48, 51, 53, 5960, 82, **116-118**, 126, **141-143**, 152-154
Papain, 24, 44
Papaya, 14, 44, 56
Passion fruits, 21, 43, 49, 56, 80
Pastes, 7, 12, 14, 24, **52-54**
Pasteurising, 3, 24, 28, 46, **48-49**, 51, 60
Pawpaws -see papaya
Pearson Square, 53-54
Peas, 10, 32-33
Peaches, 10
Pears, 10
Pectin, 24, **43-44**, 53, 55
Peelers, 25, 26, 28, 30, 42, 46, 53

Peppers, 21, 35, 45, 80
pH meters, paper, 49, 139
Pickles, 3, 10, 13-14, 24, **38-42**
Pineapples, 8, 10, 14, 48, 56
Planning, production, 20, **76-84**, 101-118
Plantains, 8, 80
Plastic - films, 25, 31, 36-37, **117-118**, 143
- pots, 37, 117
- sealers, 25
Pomegranate, 21
Potable water, 104, 137
Potato, 36
Preservation Index, 38-40, 45
Problems of food processing, 1, 15, 17, 20, 82
Processing
- blanching, 3, 29, 30-33, 110, 113
- bottling, 3, 24, 27-28, 48
- canning, 3, 24, 26, 29
- concentrating3, 12, 24, 42, 46, 78
- distilling, 3, 24, 56, **61**, 114-115
- drying, 3, 11, 24, 29-37, 108, **111-112**
- fermenting, 3, 24, 39, **56-61**, 114-115
- filling, 25, 28, 31, 35, 39, 42, 46, 48-50, 51, 53, 59, 60, **115-117**, 142-143
- for home consumption, 11-17
- frying, 3, 24, 25-26
- packaging, 25, 28, 31, **35-37**, 39, 42, 46, 48, 51, 53, 59, 82, **116-118**, 126, 141-143, 152-154
- pasteurising, 3, 24, 28, 46, 48, 51, 60
- pickling, 3, 10, 13-14, 24, **38-42**
- pulping, 13, 24, 30, 45-49, 51, 53, 59
- room - see buildings
- safety, 162
- salting, 3, 13, 38-42,
- site, 101
- sulphuring/sulphiting, 29-30, **32**, 138
- syruping, 33-34
Production planning, **76-84**, 101-118, 125-127
Products
- beer, 12
- bottled fruit, 24, **26-29**, 48,
- chutney, 7, 12, 24, **38-39**, 67-72, 76-84, 90
- cordial, 24, **50-51**, 96
- dried fruits and vegetables, 2, 7, 24, **29-37**, 110-112
- fried fruit, 24-26
- fruit cheese, 24, 52-54
- jam, 24, **52-55**, 66, 96
- jelly, 24, **52-54**, 96
- juice, 24, **47-50**, 54, 58, 96, 113
- marmalade, 24, 36, **52-55**, 96, 138
- nectar, 24, 96
- papain, 24, 44
- paste, 7, 12, 14, 24, **52-54**
- pectin, 24, **43-44**, 53-55
- pickles , 3, 10, 13-14, 24, **38-42**
- promotion, 149-150
- puree, 24, **52-54**
- quality assessment, 64-67, 130-145
- salted vegetables, 24, **45**, 65
- sauce, 13, 24, **45-46**, 65
- spirits, 24, **56, 61**
- squash, 24, **50-51**, 96
- syrup, 12, 14, 24, 28, **33-34**, 38, 50, 53
- vinegar, 14, 24, 39, 42, 46, 52, **56, 60**, 78
- wine, 10, 12, 24, **56-59**
Profit, 63, 121-123
Profit and loss statement, 158
Promotion of products, 149-150
Pulps, 13, 24, 30, 45-49, 51, 53, 59
Pulper finisher, **45**, 48
Purees, 24, **52-54**

Q

Quality assurance, 19, 25, 28, 30-31, 39, 42-43, 46, 48, 51, 53, 59-60, 76, **130-145**

R

Raisins, 36

Raw materials, 13, **76-78**, 119-121, 127, 134-137, 159-160
Reasons for food processing, 6, 9, 19-20
Records
- financial, 85-90, **155-159**
- legal, 93-99
- production, 157, **159-160**
- sales, 156-157
Refractometer, 28, 30, **38**, 39, 42, 46, 51, 53, 138, 140
Registration of businesses, 92
Relative humidity- see humidity,
Retailers, 16, **121-123**

S

Safety of foods, 130-145
Sales records - see records
Salting, 3, 13-14, 24, 38-42
Sanitation, 97-99, 104-106, 161-162
Sauces, 13-14, 24, **45-46**, 65
Scale of production, 22
Scales, weighing, 19, 25, 30, 39, 42, 46, 48, 51, 59, **110**
Seasonality, 20, 21
Slicers, 25, 28, 30, 39, 42, 43, 48, 138
Sodium benzoate, 45, 47, 50, 53
Sodium metabisulphite, 25, **32**, 49
Solar dryers, 11, 35
Sorting, 3, 25, 28, 30, 39, 42, 46, 48
Spices, 14, 38-42, 46, 96
Spirits, 24, **56**, **61**
Spoilage, 9, 20
Squashes, 24, **50-51**, 96
Staffing levels, 82-84
Start-up costs, 85-86
Steamers, 30, 113
Sterilisation, 27
Storage, 14-17, 25, 28, 31, 39, 42, 46, 48, 51, 53, 59, **141-143**, 157
Sulphiting/sulphuring, 29, 30, **32**, 138
Sun-drying11, **34-35**, 138
Support institutions, 167-171
SWOT analysis, 72-75
Syrups, 12, 14, 24, 28, **33-34**, 38, 50, 51, 53-54

T

Tanks, water, 104
Thermometers, 27, 54, 134, 139
Tomatoes, 8, 10, 12, 45-46, 52, 66, 96
Traders, 119-121
Traditional processing, 7, 57-58

U

Ultra-violet light, 105-106
Unincorporated associations, 9, 93

V

Variable costs, 87-90
Vegetarianism, 7, 148
Vinegar, 14, 24, 38-39, 42, 46, 52, **56**, 60, 78
Vitamins, 7-8

W

Washing, foods, 3, 25, 28, 30, 39, 42-43, 46, 48
Wastes, 17, 98, 101, 105-106, 137
Water - disinfection, 104-106
Weighing scales - see scales
Wines, 10, 12, 24, **56-59**
Working capital, 85

X,Y,Z

Yeasts, 56-59